REFERENCE
IN-LIBRARY USE ONLY

FLORA
AND FAUNA
of the
CIVIL
WAR

FLORA
AND FAUNA
of the
CIVIL
WAR

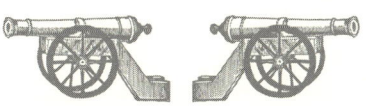

AN ENVIRONMENTAL
REFERENCE GUIDE

 **KELBY OUCHLEY**

LOUISIANA STATE UNIVERSITY PRESS
BATON ROUGE

Published by Louisiana State University Press
Copyright © 2010 by Louisiana State University Press
All rights reserved
Manufactured in the United States of America
First printing

Designer: Laura Roubique Gleason
Typeface: Arno Pro
Printer: McNaughton & Gunn, Inc.
Binder: John H. Dekker and Sons

Library of Congress Cataloging-in-Publication Data

Ouchley, Kelby, 1951–
  Flora and fauna of the Civil War : an environmental reference guide / Kelby Ouchley.
    p. cm.
  Includes bibliographical references and index.
  ISBN 978-0-8071-3688-1 (cloth : alk. paper) 1. United States—History—Civil War,
1861–1865—Environmental aspects. 2. Ecology—United States—History—19th
century. 3. Natural history—Middle Atlantic States. 4. Natural history—Southern
States. 5. Animals—Middle Atlantic States—History—19th century. 6. Animals—
Southern States—History—19th century. 7. Plants—Middle Atlantic States—
History—19th century. 8. Plants—Southern States—History—19th century.
9. Middle Atlantic States—Environmental conditions. 10. Southern States—
Environmental conditions. I. Title.
  E468.9.O93 2010
  578.0973'09034—dc22
                        2010003733

To my brother Keith,
fellow time traveler through the swamps and across the plains
of historical ecology,

and to the memory of our great-great-grandfathers who walked the walk:
Private Dallas Ouchley, Company G, 31st Louisiana Infantry
Lieutenant Henry W. Reddick, Company E, 1st Florida Volunteer Infantry

CONTENTS

Part II: Fauna

Illustrations follow page 108.

ACKNOWLEDGMENTS

Three unforgettable mentors shoved me off into a vocation of natural history. Dr. Neil Douglas, Dr. R. Dale Thomas, and the late Dr. Tom Kee, biology professors at what was then known as Northeast Louisiana University, first had to teach me how much I didn't know before they could inject me with an education proper. Their pedagogy emphasized taxonomy, be it of freshwater darters, adder's tongue ferns, or flycatchers, and the "survival of the fittest" theory applied as well to their students. I can't thank them enough.

This book would not have been possible without the cheerful, persistent efforts of the reference librarians at the main branch of the Ouachita Parish Public Library. Their diligence in tracking down and procuring research materials throughout the United States is a credit to their profession.

Rand Dotson, Lee Sioles, Catherine Kadair, and Kate Barton have made my experience with LSU Press nothing but positive. The high caliber of this publisher is a reflection of their quality of work. Likewise, freelance copyeditor Jane McGarry sanded the rough edges from my manuscript like the pro that she is.

Finally, my wife Amy, a naturalist with mystical insights, provides me daily doses of inspirational nurturing. I stay near her too in hopes of absorbing some of her radiant creativity.

FLORA AND FAUNA
of the
CIVIL
WAR

INTRODUCTION

Why write a book on the wild flora and fauna of the Civil War? One historian notes, "more than three decades of accumulated literature in environmental history barely touch the Civil War."[1] That surprisingly few efforts have been made to address any aspect of the environmental history of an epoch that has otherwise spawned millions of written words is only a partial answer. Dissecting the affairs of the United States from 1861 to 1865 reveals new issues relevant to the natural history, including the flora and fauna, of America. Humans impacted plants and animals on an unprecedented scale as millions of soldiers and refugees tramped across the landscape foraging and waging war. Likewise, at a new magnitude, wild plants and animals impacted people in the form of barriers, disease vectors, medicines, food, shelter, and raw products necessary to implement war. The Civil War exposed some people to particular flora and fauna for the first time or in new ways. For example, before the war, many northern soldiers had never seen Spanish moss, and although bats were familiar animals, no one knew that bat guano would become a critical component of gunpowder for the South. Significantly (at least for historians) and because of an increasing rate of literacy in the country, large numbers of people from all walks of life wrote about their encounters with the natural world as never before. The war generated countless letters, diaries, and journals, many of which contain accounts of plants and animals, some as comments in passing and others as part of a description of a noteworthy life event. Collectively, these first-person anecdotes reveal insights on environmental perspectives of the era.

This book, then, a blend of traditional and natural history, is an attempt to portray the roles and uses of many wild plants and animals during the American Civil War. A second, equal goal is to examine how people, soldiers and citizens alike, thought about wild flora and fauna in a time of epic historical events, as recounted in their own words.

The mixed marriage of disciplines encountered in this study has produced an offspring similar to both but necessarily lacking in some common characteristics of each. For example, the species accounts do not emphasize detailed and widely available facts about life history. Also, the species are not listed according to a taxonomic order. Scientific nomenclature is used sparingly and typically as a first reference to a plant or animal at the genus or species level. General data pertaining to ecology, conservation, taxonomy, range and habitat, and other natural history aspects are given if they are thought to contribute to understanding the flora and fauna's context during the Civil War. Only wild plants and animals are considered. Important domesticated species such as cotton, horses, and mules have been thoroughly examined elsewhere. The geographic focus of this book is the area where most of the Civil War was fought—roughly, the region east of the 100th meridian and south of the 40th parallel, and including the coastal waters of the Atlantic seaboard and Gulf of Mexico.

A major part of this book consists of anecdotes from the Civil War era pertaining to wild flora and fauna. In most but not all cases, the anecdotes were gathered from contemporary diaries, letters, and journals. I agree with the observation of Colonel Thomas W. Higginson, leader of the first African American regiment (1st South Carolina) of the war: "There is nothing like a diary for freshness."[2] Diaries, letters, and journals capture a freshness of thought, clarity, and accuracy that is sometimes lost in memoirs, reminiscences, and other sources penned long after the fact. The passages were written by Union and Confederate soldiers of all ranks in all theaters of the war, as well as by politicians, nurses, physicians, wealthy socialites, destitute wives, and foreign observers. With diverse backgrounds and life experiences, they wrote of flora and fauna from different perspectives. Barely literate farm boys familiar only with their county of birth and widely traveled, college-educated "gentlemen" noted natural phenomena with equal enthusiasm but dissimilar points of view. To aid the reader in gauging the context of the anecdotes, the date and location of the account, along with the rank and unit of soldiers and sailors, are given when known. The spelling and punctuation within the passages have been preserved as found but in some cases have been altered a bit by earlier editors for comprehension. Three dots (. . .) indicate that some part of a sentence has been omitted for conciseness. Brackets [ ] within a quote contain my words. Latin scientific names are in italics.

The anecdotes can be broadly categorized according to why the authors mentioned flora or fauna in their accounts. Some writers recorded wild plants and animals as descriptive components of the habitats to which they were exposed.

Private William G. Bentley, 104th Ohio Volunteer Infantry, in a letter to his family from Scott County, Tennessee, on Aug. 28, 1863: "I never saw such a variety of trees and bushes. Black, white, red & chestnut oak, chestnut, hickory, poplar, white & yellow pine, cottonwood, persimmon, chincopin, laurel, etc."[3]

Some referred to them as food or other utilitarian items.

Lieutenant Colonel William H. A. Speer, 28th North Carolina Regiment, in a letter to his mother from near Fredericksburg, Virginia, on April 6, 1863: "Tell father I have been living on shad for four days & have five nice ones salted up—how I wish he had them. Our men ketch hundreds of them here in the River."[4]

Frederic A. James, U.S. Navy, while a prisoner of war at Salisbury, North Carolina, on Feb. 27, 1864: "We bought a new broom made of a kind of wild grass called 'broom sedge,' today,—price fifty cents C.S. currency. . . . It is not very tough or durable, but is a much better article for a clean sweep than the stiff birch 'scrub' brooms furnished us by the 'C.S.' 'Broom Sedge' looks much like oat straw, but is stouter, as instead of being hollow it has a kind of pith like corn stalks."[5]

Others cited flora and fauna as objects of curiosity or as figures of speech.

James T. Ayers, 129th Illinois Volunteers, at sea near Beaufort, South Carolina, on Jan. 12, 1865: "we saw on Saturday hundreds of great Sea Horses [porpoises] rising fare up in sight in there Sportive play Running after Around and even Passing us. Thease sea monsters are verry Large and are queer Looking Customers and while they were Sporting and Raring as they did the Sailors said we would have A storm and So it was."[6]

President Abraham Lincoln in a letter from Washington, D.C., to Cuthbert Bullitt on July 28, 1862: "What would you do in my position? Would you drop the war where it is? Or, would you prosecute it in future, with elder-stalk squirts, charged with rose water?"[7]

Some passages convey factual natural history information. Others reveal applied uses of wild plants and animals. Whether flora and fauna were the focus of an anecdote or an insignificant sidelight, the sketches often expose scenes and moods beyond the natural history or practical realms that facilitate a deeper understanding of the conflict regarding the individuals who were at ground level. Perhaps the real value in attempting to blend natural and traditional history lies in this potential.

THE CIVIL WAR SETTING

Flora

Before humans migrated to North America, forests likely covered 650 million acres east of the Great Plains in what is now the United States.[1] Anthropomorphic change began with the first Native Americans and their use of fire to alter the landscape. When Europeans reached the continent, the primeval forests had already been modified, although the extent of the change remains poorly understood. European settlers accelerated the pace drastically with new technologies, especially those related to clearing land for agriculture (e.g., iron tools and draft animals).

Increasingly, farms replaced forests for two hundred years up until the time of the Civil War. Forests were cleared to plant corn, cotton, rice, tobacco, and vegetables. By 1860 there were an estimated 81.3 million acres of agricultural lands in the North. States with the most farmlands were New York (14.5 million), Ohio (12.6 million), and Pennsylvania (10.5 million). At this time there was an estimated 73.4 million acres of agricultural lands in the South, including the border states of Kentucky and Missouri. Southern states with the most farming were Virginia (11.9 million acres), Georgia (8.4 million), and Tennessee (6.9 million).[2] Some farmlands, such as those in the highly erodible Piedmont region, had already been abandoned and reverted to old-field pines by the time of the Civil War.[3] Associated with most farms, domestic or feral livestock impacted native vegetation in many areas. Hogs, in particular, with their omnivorous diets and rooting habits, are agents of change in any natural system. The 1850 census recorded more than two hogs for every person in Alabama, Florida, Georgia, Mississippi, and North Carolina.[4]

The developing nation made additional demands on natural forests. Lumber was sought for buildings, wagons, crossties, and barrels. At the beginning of the Civil War, North Carolina's pine forests were the country's center of production for naval stores. Wood for ship planking, spars, and masts, and resin products such as pitch and tar for waterproofing, preserving, and turpentine were in great need.[5] Fuel wood fed the boilers of steamboats and locomotives, as well as fireplaces for domestic heating and cooking. Charcoal made from various species of cordwood fueled the fledgling iron industry. Hardwood ashes were used in the manufacture of potash—a basic ingredient in soft soap and gunpowder, and hardwood bark was necessary to tan leather.[6] As the following passage shows, the loss of forests in some places was lamented as early as 1849: "the time is not distant when public attention must be drawn to the planting of forest trees in the country . . . we have destroyed our forests . . . and posterity must face the consequences."[7]

However, in spite of extensive human impacts, most of the area directly affected by the Civil War was still forested in 1861. Indeed, sizable tracts of virgin forests remained, and large-scale commercial logging did not begin until after the war. Deciduous hardwoods comprising several species of oaks (*Quercus* spp.) and hickories (*Carya* spp.), chestnut (*Castanea dentata*), yellow-poplar (*Lireodendron tulipifera*), and sweetgum (*Liquidambar styraciflua*) dominated the canopies of eastern interior forests. The coastal regions of the Carolinas and Deep South were blanketed with vast coniferous forests of longleaf (*Pinus palustris*), loblolly (*P. taeda*), and slash (*P. elliottii*) pines. Beneath the larger trees floristic diversity was great, as numerous midstory and shrub species grew alongside a host of perennial and annual wildflowers, vines, grasses, and ferns. Throughout and on the periphery of the general forested landscape, unique habitat types in the form of prairies, marshes, savannahs, bogs, brakes, glades, and balds existed on a much smaller scale in response to special environmental conditions. Civil War forests were not the vast unbroken stands of giant, primeval, old-growth trees depicted by early nineteenth-century romantic writers. Rather they were likely a mosaic of types, ages, and sizes resulting from many factors including various site characteristics (e.g., soil fertility, moisture, slope, latitude, etc.) and past disturbance (e.g., human- or lightning-induced fire, hurricanes and other storms, destructive insects, and disease).

Native plant data relevant to specific sites such as battlefields during the war are scarce. During the research for this book, I contacted the superintendents of all Civil War national military parks seeking information on flora and fauna of their areas at the time of the conflict. The lack of information was surprising. If any information was available, it usually consisted of broad, general descriptions of forested areas. At best, period descriptions and maps denote "heavy woods," "thicket," "cane," "swamp," or "cedar brake." Rare exceptions occur, but only for woody species. An example is the well-researched map *Historical Land Cover/Use Classification of Pea Ridge National Military Park.*[8]

Fauna

The first Native Americans also impacted the fauna of North America. Paleolithic hunters sought now-extinct megafauna such as mammoths, mastodons, and giant bison. The efficiency of skilled hunters may have significantly contributed to the extinction of some species already stressed in an era of global climate change. Europeans brought firearms and continued the pressure, particularly on game animals. By 1850 most large game mammals had vanished from the Northeast.[9]

Before the Civil War, laws to protect wild animals were uncommon, rarely enforced, and generally ineffective. Any species with commercial value was subject to market hunting, and predators were pursued opportunistically. Rhode Island enacted one of the first game laws in the United States in 1646 by designating a closed season for deer. Virginia prohibited the killing of female deer in 1738, and New York barred hunting of deer with hounds in 1788. Some southern states established deer seasons. Rhode Island prohibited spring shooting of waterfowl in 1846, but it was not until eleven years before the war that laws were enacted to protect songbirds, in Connecticut and New Jersey.[10]

At the outbreak of the Civil War, the types of fauna found in the eastern United States were very similar to those there today. A broad diversity of mammals, birds, fishes, reptiles, amphibians, and invertebrates lived in a variety of habitat types. Populations varied naturally and as a result of human activities. Deer and bear were uncommon or eliminated in some areas along the eastern seaboard but thrived in remote swamps and mountainous areas. All mammalian predators were pursued relentlessly as threats to progress. Rabbits, quail, and other species adapted to the forest edge flourished around farms yet to be subjected to pesticides. Waterfowl populations were so large as to seem uncountable, especially on their wintering grounds, consisting of pristine wetlands. Songbirds and birds of prey, although persecuted, were abundant. Fish and other aquatic species thrived in clean waters free of pollution (with a few small exceptions).

Humans and Their Attitudes

Early American pioneers maintained European attitudes toward nature, considering it an entity to be conquered, civilized, and rid of competing wild beasts as necessary. The theory of Manifest Destiny reflected a theological belief that settlers were divinely appointed to "use" the earth for the enhancement of civilization, no holds barred. Such attitudes eventually led to the decimation of many Native American cultures and the extinction or near extinction of several animals.

Barely tempering this prevalent thought in the 1830s and 1840s, the poets Byron and Wordsworth and artists associated with the Hudson River School of painting championed a "romantic attitude" that promoted nature for its aesthetic values. With his publication of *Walden* in 1854, Henry David Thoreau argued not for sentimentality but for the wisdom of seeking God in nature. His position contained the embryonic thoughts of what became the "preservation attitude" after the Civil War. Of the hundreds of personal anecdotes gathered for this book, those expressing aesthetic or religious sentiments toward wild flora and fauna are uncommon.

The South, where most of the actual fighting of the Civil War occurred, was essentially rural. Of the hundred largest cities in the United States in 1860, only fourteen were in the South. Those in the top twenty-five included New Orleans (sixth), Louisville (twelfth), Charleston (twenty-second), and Richmond (twenty-fifth). Atlanta was ninety-ninth in size.[11] Census data reveal that the free population of the entire United States in 1860 (excluding the far western states and territories) was 26,770,614. Of this number, the free population of the thirteen states in the Confederacy was 7,479,504.[12] Generalities drawn from these figures include these: On average, Confederate soldiers and other participants from the South likely had more experience with nature and its components than their northern counterparts, and a large number of northern soldiers from urban areas had minimal experiences with the natural world. In fact, these assumptions are confirmed in journals, diaries, and letters from both sides of the conflict.

A sense of the prevailing attitudes of soldiers toward wild flora and fauna can be acquired by a rough analysis of the anecdotes by category, as presented in the table below. The "Habitat Component" category was used when the writer mentioned wild plants or animals primarily as a part of the habitat to which he was exposed. The "Curiosity" category contains anecdotes that refer to flora and fauna encountered more as objects of curiosity or novelty. "Utilitarian Item" includes such miscellaneous things as utensils, shelter, bedding, firewood, and implements of war (e.g., abatis). "Food," "Medicine," and "Pest" categories are self-explanatory. Research for this book yielded more sources from northern soldiers and noncombatants than from those with Confederate sympathies. This fact is reflected in the number of anecdotes suitable for the simple analysis—359 Union versus 148 Confederate.

Frequency of comments concerning wild flora and fauna in the letters, diaries, and journals of Union and Confederate soldiers

	Flora and Fauna		Flora		Fauna	
	Union (%)	Confed. (%)	Union (%)	Confed. (%)	Union (%)	Confed. (%)
Habitat Component	27	17	36	29	17	4
Food	29	43	22	32	36	57
Utilitarian Item	12	12	24	20	1	1
Curiosity	18	13	10	6	26	22
Medicine	4	8	8	13	0	0
Pest	10	7	0	0	20	16
Total (%)	100	100	100	100	100	100

Because of the great social, cultural, educational, and experience diversity among Civil War participants on both sides, one must be careful not to broad-brush paint a particular bloc with unique characteristics. However, analysis of the anecdotes prompts interesting deductions. Union and Confederate soldiers mentioned wild flora and fauna combined most often in the context of food; second, as a component of the habitat. As a percentage of their total comments, though, Confederate soldiers referred to food much more often than did Union soldiers (43 percent versus 29 percent). Also as a percentage of their total anecdotes about wild plants and animals, Confederate soldiers wrote of flora as medicine twice as often as Union soldiers (8 percent versus 4 percent). These figures parallel generally held beliefs that the southern armies often suffered for want of traditional rations and medicine more than their northern counterparts. As this was the case, Union soldiers would not have needed to regard wild flora and fauna in the context of food or medicine as often as southern soldiers. Some consideration must also be given to the likelihood that Confederate soldiers were more familiar with the value of wild plants and animals as food and medicine on their home turf. Union soldiers mentioned wild flora and fauna as habitat components and curiosities more frequently than Confederate soldiers. This no doubt was at least somewhat due to the fact that northern soldiers were more likely to be farther from their homes and thus encountering unfamiliar species more frequently. Southerners, on the other hand, were mostly rural, and many experienced the war in familiar ecosystems in which the flora and fauna were well-known components. Under such conditions, wild plants and animals were not as noteworthy unless they served an important practical purpose. Of course, exceptions abound.

The patterns are similar when anecdotes about flora and fauna are separated. Confederate soldiers mentioned flora most often in the context of food, but for Union soldiers the food category ranked third. The category rankings for fauna were the same for Union and Confederate soldiers, although southern soldiers mentioned fauna as food at a much higher frequency (57 percent versus 36 percent). Union soldiers also wrote of fauna as habitat components and curiosities at a much higher rate than Confederates.

Looking closely at the food category reveals that both Union and Confederate soldiers mentioned wild fauna in the context of food much more often than they did wild flora. This interesting fact may be in part the result of lingering societal attitudes about men as hunters and women as gatherers. That desirable wild plant foods were available only for relatively short periods during the year may be another factor. When Union soldiers did write of wild plants as food, they mentioned blackberries more than twice as often as any other plant. Chestnuts and chinquapins and grapes were the most common wild flora mentioned as food by Confederate soldiers. Regarding fauna, Union soldiers recorded fish as food most often, followed by wild birds of various species. Fish also ranked highest with Confederate soldiers, but birds placed third just behind squirrels. The frequent mention of fish was likely due to the availability of fish in all of the many water bodies within the Civil War arena the year round, the relative ease of acquiring fish, and the high value placed on fish as a desirable food source. Hunting and consuming squirrels was more a rural southern than a northern tradition, then as today.

The sample size of noncombatants is too small to analyze in a meaningful way by affiliation, gender, or race. General trends followed those of soldiers with the same political affinity. Both northern and southern women mentioned wild flora twice as often as fauna. Southern women seemed especially occupied with food, medicine, and utilitarian items.

PART I

FLORA

We elms of Malvern Hill
Remember every thing;
But sap the twig will fill:
Wag the world how it will,
Leaves must be green in Spring.

—HERMAN MELVILLE,
"Malvern Hill," July 1862

INTRODUCTION

Flora is often defined as plants of a particular region or period. Most floras are comprehensive, systematic accounts. The classification of plants is constantly evolving, and some groups that were once considered plants such as algae and fungi are now placed in separate categories. Vascular plants are those that have specialized tissues for conducting water, minerals, and nutrients. Ferns, conifers, and true flowering plants are in this group. Conifers, such as pines (*Pinus* spp.) and baldcypress (*Taxodium distichum*), are sorted by having seeds not enclosed in an ovary. True flowering plants produce flowers and fruit and include most common plants such as deciduous trees and shrubs, vines, annual and perennial wildflowers, herbs, and grasses. For purposes of this book, only conifers and flowering plants are considered. Lesser groups were of little practical importance and rarely mentioned during the Civil War.

Plants have been used throughout history as medicines, and this use may have reached a high point during the Civil War. A medical sect of the period known as "Botanics" favored the use of roots and herbs for treatment.[1] *Tisanes* was a common term for "household remedies made from flowers, bark, red pepper, and roots."[2] Prior to the war most pharmaceutical factories were in the North, which forced the makeshift southern labs in wartime to lean heavily on native plants for substitute drugs. Confederate surgeon general Samuel P. Moore was a firm believer in the use of plants and listed sixty-five species in his *Standard Supply Table of the Indigenous Remedies for Field Service and the Sick in General Hospitals.*[3] He also encouraged Francis P. Porcher's monumental and widely quoted treatise, *Resources of the Southern Fields and Forests, Medical, Economical, and Agricultural,* a major medical botany of the war. The use of plants was rationalized by a widespread belief of the era that each area contained plants that could be used to treat diseases common to that area.[4] As a large body of detailed work exists on medicine and pharmacopoeia during the Civil War, here the use of plants for those purposes is treated broadly.

Civil War–era writers frequently wrote about other practical uses of plants, such as food, fuel, shelter, and lighting:

Private William R. Stilwell, 53rd Georgia Volunteers, in a letter to his wife from near Fredericksburg, Virginia, on May 28, 1863: "We make greens out of the polk [poke] and other weeds. I don't like it much. General Semmes eats it most every day; I [facetiously] say starve the Southern army—you might as well try to starve a black hog in the piney woods."[5]

Private Robert A. Moore, 17th Mississippi Regiment, on Goose Creek, Virginia, on Nov. 4, 1861: "Had me chopping wood to-day. My axe found any quantity of black haws."[6]

Private Henry R. Berkeley, Amherst Battery in the Army of Northern Virginia, establishing winter quarters in Louisa County, Virginia, in January 1863: "Our whole time taken up in work on our stables. We built large pens out of large pine logs and covered them with plank. We had pressed a sawmill into use and got on quite rapidly."[7]

Refugee Kate Stone in Tyler, Texas, on Feb. 15, 1865: "the servants make such pretty candles now. . . . They boil a species of cactus in the tallow, and the candles are partly transparent and brittle and give an excellent clear light."[8]

Plants were also the subject of a variety of anecdotes on different issues ranging from danger to celebrities:

Lieutenant John Q. A. Campbell, 5th Iowa Infantry, near Memphis, Tennessee, on Feb. 19, 1863: "The wind blew a large tree down in camp today, crushing two tents and injuring four men, in Companies K and I—one severely."[9]

Sargeant John Westervelt, 1st New York Volunteer Engineer Corps, on the Appomattox River, Virginia, on March 5, 1865: "While here I visited the celebrated tree, called the Pochantas Oak. It is said to be the identical tree under which Pochantas saved the life of Captn John Smith from the vengeance of her father King Powhattan. It stands on the brink of the rocky precipice some fifty or sixty feet above the waters of the Appomattox. . . . The tree, though I have seen larger, is the most symmetrical I ever saw. A sign is nailed on it with the quotation, Woodman spare this tree, Notwithstanding this it has been considerable cut and hewn and I could not resist the temptation of bringing away a piece of it myself."[10]

Often plants were mentioned as components of a habitat, either in the focus of a passage or as an incidental remark. No ecosystem in the Civil War arena, from the western frontier to the Wilderness Battlefield to the Florida Keys, was overlooked:

Rev. Francis Springer, chaplain, 10th Illinois Cavalry, at Lindsay's Prairie, northwestern Arkansas, on August 17, 1863: "There are here & there small groves of dwarf elms, persimmons, & cherry plums with grape vines & blackberry bushes. These cherry plums are the most beautiful & delicious looking wild fruit I have seen. The fruit is arranged in clusters of 5 to 8 plums in a bunch & these bunches are very close to each other, making a fine appearance, especially as the color of each cherry is a bright glossy red or yellow."[11]

Major William J. Bolton, 51st Pennsylvania Volunteers, near Warrenton, Virginia, on May 6, 1864: "It [Battle of the Wilderness] was fought in the midst of dense thickets of second-growth underbrush, evergreens, pines, sweet-gums, scrub-oak, and ceder, rendering the use of artillery impossible, and compelling the opposing lines to approach very near each other in order to see their opponents."[12]

John Hay, assistant secretary to President Lincoln, at Key West, Florida, on March 8, 1864: "The town looks more like the tropics than anything I have yet seen. The cocoanut trees are the special feature of the streets: every yard is distinguished by their huge fanlike branches."[13]

This book lists plants alphabetically by common name. The heading "Herbs" is an exception and includes a group of plants with common uses.

ASH

"If ash is out before the oak, twill be a summer of fire and smoke, but if oak is out before the ash, twill be a summer of wet and splash."[1] Adages involving the behavior of flora and fauna as a predictor of weather were common during the Civil War. Ashes (*Fraxinus* spp.) are members of the olive family. The eight or nine species in the eastern United States are medium to large trees easy to identify as a group by their opposite, compound leaves. They grow in a variety of habitats from wet (pumpkin ash, *F. profunda*) to dry (white ash, *F. americana*) sites. Their winged seeds, known as samaras, are eaten by birds and mammals. White ash and green ash (*F. pennsylvanica*) were important timber trees and yielded construction lumber, flooring, furniture, farm implements, wagon and carriage components, barrels, tool handles, oars, and drum shells. The smaller species were used to manufacture baskets, chair bottoms and splints, and hoops. Bark from blue ash (*F. quadrangulata*) produced a blue dye.[2]

A tonic made from shredded ash bark was used as an astringent, diuretic, and to treat arthritis, fever sores, and constipation. Snakebite was treated with tea made from ash buds.[3] This therapy may have roots in another superstition of the era: leaves of white ash "are said to be so highly offensive to the rattlesnake that that formidable reptile is never found on land where it grows; and it is the practice of hunters and others having occasion to traverse the woods in the summer months to stuff their boots or shoes with white ash leaves as a preventive of the bite of the rattlesnake."[4]

All species of ash have heavy, dense wood with high BTU values, an excellent quality for firewood.

Private Isaac Jackson, 83rd Ohio Volunteer Infantry, near Paris, Kentucky, on Oct. 21, 1862: "And after we broke ranks we went to the fence to get some wood . . . and we got a load and took it to the fire. . . . In a few moments the camp fires were brightly burning. They were very nice ash rails and made good fire."[5]

Private Robert Patrick, 4th Louisiana Infantry, at Port Hudson, Louisiana, on Dec. 9, 1862: "Our chimney makes my tent very comfortable, but when the fire goes out, I get quite cold. . . . We have splendid wood to burn. I had a large ash tree cut down and split up into wood of the right length to suit my fire-place."[6]

Fringe tree (*Chionanthus virginicus*), also called grancy gray beard among other names, is a shrub or small tree in the ash family. Prized for its showy white, fringelike flowers, fringe tree has been cultivated since 1736 and was widely planted in yards by the time of the Civil War.[7]

Sarah Wadley, daughter of the supervisor of Confederate railroads, near Trenton, Louisiana, on April 29, 1864: "This morning immediately after breakfast we all took a little walk to gather flowers, got a quantity of the beautiful graceful blossoms of the 'Old man's beard' tree."[8]

BALDCYPRESS

Nothing characterizes a southern swamp more than a giant moss-draped baldcypress (*Taxodium distichum*) standing knee-deep in a backwater slough. Commonly known as "cypress," these survivors of ancient life forms once found across North America and Europe are now greatly restricted in range. In the United States they are native to river bottoms and swamps in the South and along the eastern seaboard north to Delaware. Baldcypress trees once grew to seventeen feet in diameter and one hundred forty feet in height. They were the largest trees in the eastern United States and lived to be four hundred to six hundred years old. A few were estimated to be more than one thousand years old. Some animals eat baldcypress seeds occasionally, but the greatest wildlife value of the tree is as den sites for species as varied as wood ducks and black bears. Humans have long sought baldcypress for its wood, which is easy to work and renowned for its resistance to decay. Before and during the Civil War the baldcypress logging industry was small, and the logging methods were primitive and depended on spring floodwaters to move logs to sawmills. The result was that only the most accessible trees along waterways were harvested. Massive exploitation of baldcypress began shortly after the war and resulted in near total elimination of the country's virgin stands.

Settlers copied the Native American tradition of using baldcypress for dugout canoes and expanded the use into shipbuilding. Many Civil War vessels, especially those used on inland waters, had baldcypress hulls.[1] Old trees with heartwood having a reddish hue yielded the most durable lumber, which was also used for railroad ties, cooperage, furniture, shingles, pilings, coffins, and general construction. In 1863, Federal troops occupying Natchez seized the entire baldcypress stock of a local sawmill, two million board feet valued at $62,000, for use by the Federal army and a "Contraband Camp." Earlier in the year, Confederate troops confiscated 673 baldcypress timbers from the same sawmill to build a boom across the Yazoo River at Snyder's Bluff as an obstacle to Federal gunboats.[2] Southern defenders at Fort Jackson below New Orleans built a raft of baldcypress logs and secured it in the river to impede northern gunboats and keep them in the fort's sector of fire.[3] Medicinally, a resin obtained from baldcypress seed cones was used to treat skin cuts and wounds.[4]

Baldcypress swamps often evoked emotions of gloom and dread among writers of the period. The unique rootlike growths commonly known as "cypress knees" were poorly understood curiosities, important in tree respiration and as storage areas for starches, not as "germs of the future tree."

Sarah Wadley, daughter of the supervisor of Confederate railroads, traveling as a refugee through the swamps of northeastern Louisiana on Oct. 3, 1863: "The shore of the bayou near where we sat was grown up in places with cypress knees of a thousand grotesque shapes, some groups of them if inverted would have looked just like great brown icicles. The growth of the cypress tree is so singular and what is strange in this day of popular information, I have never seen any notice of it in books of travel. . . . These knees, as they are called, are the germs of the future tree, they grow up like stumps . . . not a bud or sign of a leaf is seen anywhere about them. . . . Mr. Duvall says they grow thus quite tall, sometimes higher than a man, and then all at once put out branches."[5]

Major James A. Connolly, 123rd Illinois Infantry, near Louisville, Georgia, Nov. 28, 1864: "the creek proper [Rocky Comfort], as well as the swamp is filled with cypress trees; these 'Cypress Knees' grow up out of the water, are very broad at the base and taper to a point."[6]

Lieutenant John Q. A. Campbell, 5th Iowa Infantry, aboard the steamer *Von Phul* below Helena, Arkansas, on March 4, 1863: "The dark forests of cypress that live by both banks may be very inviting to beasts and reptiles, but they have very little attraction for the human eye, at this season."[7]

Sarah Wadley, daughter of the supervisor of Confederate railroads, traveling as a refugee through the Tensas Swamp of northeastern Louisiana on Sept. 30, 1863: "At one place we caught a glimpse of a lake through the forest, I rode in where I could see it better and never shall forget the scene, a stillness as of death reigned over the green water, covered for a large space near the land with the leaves of the water lily, the cypress trees hanging over the water were draped with grey moss, silent and still, waved by no breath of passing breeze."[8]

Private Alexander Downing, 11th Iowa Infantry, near Bayou Macon in northeastern Louisiana on Aug. 24, 1863: "The country here is low and heavily timbered with cypress and the ground is covered with masses of palm [palmetto] leaf."[9]

Lieutenant Charles B. Haydon, 2nd Michigan Infantry, at Hampton Roads, Virginia, on March 23, 1862: "Among the new trees which I have seen are cypress & holly, both very pretty especially the latter."[10]

Lieutenant Jacob Ritner, 25th Iowa Infantry, in a letter to his wife from near Delta, Mississippi, on Nov. 30, 1862: "The cypress swamps are a grand sight—trees almost as thick as they can stand, and from 60 to 80 feet without a limb."[11]

Lieutenant John Q. A. Campbell, 5th Iowa Infantry, near Helena, Arkansas, in a letter to the [Iowa] *Ripley Bee* on March 12, 1863, describing the Yazoo Pass expedition: "The rebels had felled large cypress trees across the pass and they had to be removed by man power. It was a heavy job but the boys had a merry time over it. At first, they were rather shy of the mud and water but they soon got used to it. The modus operandi was simple. They would tie a large cable to a tree and as many as could would take hold and walk off with the tree. Large cypress trees—measuring three feet at the butt and carrying their thickness for eighty feet were hauled off with perfect ease—large limbs snapping off like twigs before the tremendous power applied to them."[12]

 Cypress Rangers was the nickname of Company F, 9th Texas Rangers from Davis (now Cass), Morris, Titus, and Marion Counties, Texas.[13]

BEECH

Undoubtedly, Civil War soldiers joined thousands of other people throughout history in carving their names or initials in the smooth, steel-gray bark of the American beech (*Fagus grandifolia*). With this unique bark and lustrous, dark green leaves the American beech is one of the most beautiful and unmistakable canopy trees in eastern North America. Closely kin to oaks, beech can grow to four feet in diameter and 120 feet tall. They prefer moist but well-drained soils. During the Civil War, some people claimed that cotton would not grow where beech were once found because this tree depleted the soil of lime.[1] The fruits of beech are small edible nuts enclosed within a rough husk.

Various components of beech wood, bark, and leaves were used medicinally to treat fevers, vomiting, seasickness, headaches, tuberculosis, and "low spirits."[2] Beech wood is dense and strong, but the lumber must be carefully dried because of high shrinkage. In the Civil War era it was used to make barrels, crates, furniture, flooring, tool handles, saddle frames, shoe lasts, plane stock, spools, and toys.[3] Confederate surgeon Francis Porcher stated, "The leaves of beech trees, collected at autumn, in dry weather, form an admirable article for filling beds. The smell is grateful and wholesome; they do not harbor vermin, are very elastic, and may be replenished annually without cost."[4] Beech was also renowned as an excellent fuel wood.

The fruit of beech trees is eaten by many species of wildlife. In the Civil War period the now extinct passenger pigeon (*Ectopistes migratorius*) relished the abundant mast in beech forests. Rural farmers in many areas depended on these same nuts to fatten free-roaming hogs. Likewise, soldiers roasted beech nuts to supplement their diets on occasion.

Sergeant Allen M. Geer, 20th Illinois Volunteers, near Vicksburg, Mississippi, on Nov. 15, 1863: "Gathered beachnuts, which are very plenty in the woods."[5]

Sergeant Hamlin A. Coe, 19th Michigan Volunteer Infantry, near Covington, Kentucky, on Oct. 14, 1862: "We marched directly in the rear of the town and camped in a beautiful grove of beech trees. We lay down upon the ground for sleep tonight. I tell you I slept soundly upon beech roots."[6]

John S. Jackman, 9th Kentucky Infantry, near Beech Grove, Tennessee, on April 23, 1863: "The hills around here remind me of Kentucky. They are covered with such pastures and beautiful groves of beech trees, which are now leaving-out."[7]

Reuben A. Pierson, 9th Louisiana Infantry, in a letter to his sister from Camp Moore, Louisiana, on June 20, 1861: "Excuse the length of my letter. I am sitting in the woods under a large beech tree, upon a pine log about a quarter from the camp where there is not a single thing to disturb me."[8]

Lieutenant Theodore A. Dodge, 101st New York Volunteers, near Harrison's Landing, Virginia, on July 20, 1862: "Our line is picketed, as I said before, behind a swamp, and in a wood of maple and beech trees, shady and pleasant."[9]

Lieutenant Jacob Ritner, 25th Iowa Infantry, in a letter to his wife from near Helena, Arkansas, on Nov. 16, 1862: "We are camped right in the woods. The timber is very large and heavy. Right in our camp it is nearly all Beech, Poplar, Sassafras, Gum, &c."[10]

Sergeant Edwin H. Fay, Minden [Louisiana] Rangers, near LaGrange, Tennessee, in a letter to his wife on Sept. 5, 1862: "Good Bye, I have written this sitting on the ground under a beech tree on the bank of the Wolf River ¾ mile south of La Grange, Tenn. Do write me often my dearest one."[11]

BLACKBERRY

Corporal Rufus Kinsley of the 8th Vermont Regiment wrote from south Louisiana on April 1, 1863: "Co. returned to Bayou Boeuf, had little to eat but blackberries for three days."[1] Union General Sherman wrote in his memoirs, "I have known the entire skirmish line, without orders, to fight a respectable battle for the possession of some old fields full of blackberries."[2] Of all native plants mentioned in Civil War diaries, journals, and letters, the blackberry (*Rubus* spp.) was one of the most common. It was so prevalent and so widespread that there is little doubt that this wild fruit was at times an important part of soldiers' diets. The soldiers were not alone. Along with a host of bird species, bears, foxes, coyotes, mice, and box turtles are attracted to the sweet berries. Deer and rabbits browse the shoots and with others animals find refuge in the thick brambles.

Blackberries, members of the rose family, are native to Europe, Asia, and North and South America and consist of hundreds of species. Dewberries and raspberries are in this group. For thousands of year humans have used blackberries for food and medicinal purposes. Confederate surgeon Francis Porcher's medical treatise lists the plant as a powerful astringent used to treat dysentery, diarrhea, kidney stones, and snakebite. Recipes for Civil War–era blackberry wine and cordials touted "an approved liquor which cheers but not inebriates."[3]

The term *ecology* is often defined as the inter-relationship of an organism with its environment. Although not a typical example, humans in blackberry patches during the Civil War were nonetheless an ecological association.

Union Brigadier General Alpheus S. Williams in a letter to his daughters from Warrenton Junction, Virginia, on July 27, 1863: "The fields now are covered with the largest kind of blackberries, both the vine and the bush kind. We have been surfeited with them. For miles and miles in every day's march since crossing the Potomac the fields on both sides of the road have been at every halt, covered with men gathering these berries."[4]

Private John M. King, 92nd Illinois Infantry at Wartrace, Tennessee, on July 4, 1863: "Blackberries were ripe and many of the men strolled in the fields and woods gathering blackberries. Blackberries in Tennessee grow so large and plentiful that it almost seemed as though the army might march on and pick blackberries for a living. Any soldier that wanted to celebrate the Fourth of July, all he had to do was to make a noise by shooting off his gun and then sit down and eat his blackberries and hard tack for a Fourth of July dinner." **Private King wrote again at Kingston, Georgia, on June 24, 1864:** "We moved back up and stopped for a day or two on the Resaca battlefield. Blackberries were ripe, and all over the battlefield the berries were growing as if human blood had fertilized the soil. The first time I passed over the field in search of berries I passed by long rows of new dry earth where large numbers of men who had stood shoulder to shoulder in line of battle were now lying shoulder to shoulder in their silent and shallow graves."[5]

Private John F. Brobst, 25th Wisconsin Infantry Regiment, near Atlanta, Georgia, August 1864, in a letter to his future wife: "You must not think up there [in Wisconsin] that we fight down here because we are mad, for it is not the case, for we pick blackberries together [with the Confederate soldiers] and off the same bush at the same time, but we fight for fun, or rather because we can't help ourselves."[6]

Corporal Rufus Kinsley, 8th Vermont Regiment, near Des Allemands, Louisiana, on June 5, 1862: "Not much to eat but alligators and blackberries: plenty of them."[7]

Sergeant Charles B. Haydon, 2nd Michigan Infantry, near Washington on July 1, 1861: "Many of the men have taken to doctoring themselves & have in several cases cured with the juice of boiled blackberry roots a diarhea which baffled the Surgeon's skill."[8]

Confederate nurse Kate Cumming at a hospital in Chattanooga, Tennessee, on June 27, 1863: "We have been busy lately making blackberry cordial and blackberry preserves. I have made about twenty-five gallons of the cordial. I never was any place where there were such quantities of blackberries. The country people bring them in by the bushel."[9]

John S. Jackman, 9th Kentucky Infantry, near Abbeville, Mississippi, on June 22, 1862: "Marched 7 miles and camped. Went out foraging . . .

blackberries being plentiful, we lived well. Had a blackberry cobbler for dinner nearly every day, and G.A. would make stacks of pies."[10]

Assistant Surgeon Dr. Daniel M. Holt, 121st New York, near White Plains, Virginia, on July 24, 1863: "have just sent four men off to pick a bucket of blackberries of which there are hundreds of bushels growing upon our old encampments. They are mostly of the low or running kind— very large and juicy, but not so sweet or pleasant a flavor as those growing with us at home; still with sugar and milk (if we are so fortunate as to get the latter) they make a dish worthy of a place upon the table of an epicure. Large quantities of excellent wine, I am informed, is made of the juice of them, and I see no reason why its manufacture should not be profitable."[11]

Private John Westervelt, 1st New York Volunteer Engineer Corps, camped near the Appomattox River, Virginia, on July 14, 1864: "To day I pick my last berrys, they are scorched all up. I picked six qts and Burton bought some sugar and we made a sort of blackberry jam which is verry excellent."[12]

Private Theodore F. Upson, 100th Indiana Infantry Volunteers, near Atlanta on August 13, 1864: "Our boys are living on fruit diet mostly now. The blackberries are so thick in the abondoned fields that one can pick a ten quart pail full in a few minutes. The boys make puddings, pies and evry thing they can think of."[13]

Captain William J. Seymour, 1st Louisiana Brigade, near Hedgesville, Virginia, on July 21, 1863: "Our men were halted in an immense field of black berries, in which the whole Division regaled themselves. The troops declared that it was merely a foraging expedition and that Gen. Early had marched them there to draw rations of blackberries—rations of bread and meat being quite scanty in camp."[14]

Refugee Sarah Morgan Dawson near Baton Rouge, Louisiana, on May 31, 1862: "Poor Lucy picked me a dish of blackberries to await my arrival, and I was just as grateful for it, though they were eaten by some one else before I came."[15]

Sergeant John Q. A. Campbell, 5th Iowa Infantry, in northern Missis-sippi on July 3, 1862: "Some of the boys who came from Camp Clear Creek today brought news of the death of Jacob Stern of our company. He died yesterday while out gathering blackberries. He died for his country."[16]

Confederate sympathizer Myra Inman at Cleveland, Tennessee, on July 13, 1861: "Pretty day. Venie, Lizzie, Sues [a slave], Egbert and I went out after blackberries."[17]

Sergeant Rice Bull, 123rd New York Volunteer Infantry, in the trenches of north Georgia on June 21, 1864: "Dan began to notice his surround-ings and looking out in our front discovered something. Turning to me he said, 'R.C. look at those berries.' I looked, and about six feet from us but out of reach were a lot of low-growing blackberries, the bushes full of ber-ries. The temptation was too great for Dan to resist. . . . I was as hungry as Dan, so I crawled out beside him and we picked berries as fast as we could, eating as we picked. As we glanced ahead we could see more and more bushes, and as we cleaned up we crawled further and further. Being greatly interested in our work we must have become careless for suddenly a bul-let whizzed between our heads that were not more than six inches apart. It tore a hole through Dan's blouse but did not wound him. We slid back into our hole."[18]

Emeline Ritner in a letter from Mt. Pleasant, Iowa, to her husband, Captain Jacob Ritner, in the 25th Iowa Infantry in Mississippi on Aug. 7, 1863: "I went up to Jefferson last Saturday to pick blackberries and didn't get to come home till yesterday and had no chance to write while there. We went to the timber two days for berries, I canned up 18 qts., not as many as I expected to get as it was such hard work."[19]

Jedediah Hotchkiss, topographical engineer of the Army of Northern Virginia, near Gordonsville, Virginia, on July 21, 1862: "We ate our fill of blackberries; the whole army turned out and picked them, the hill sides were full of them; the General [Thomas 'Stonewall' Jackson] seemed to enjoy them [very] much."[20]

Sergeant Alexander Downing, 11th Iowa Infantry, near Corinth, Mis-sissippi, on June 27, 1862: "Blackberries are beginning to ripen and seem to be plentiful. Fresh fruit with our rations will lighten our work." **On June 2 he wrote from the same location,** "I went out about a half mile from camp to pick blackberries, and I picked a gallon of them and sold them to the hospital steward for $1.25."[21]

President Abraham Lincoln in a letter from Washington, D.C., to Richard Yates and William Butler on April 10, 1862: "I fully appreciate Gen. Pope's splendid achievements with their invaluable results; but you must know that Major Generalships in the Regular Army, are not as plenty as blackberries."[22]

← Blackberry Picker—"A Straggler, i.e. one who leaves his line under the pretext of gathering foodstuffs, preferring the risk of imprisonment to that of being maimed or killed in an engagement."[23]

CANE

President Theodore Roosevelt aptly described a Louisiana canebrake: "The canebrakes stretch along the slight rises of ground, often extending for miles, forming one of the most striking and interesting features of the country. They choke out other growths, the feathery graceful canes standing in ranks, tall, slender, serried, each but a few inches from his brother, and springing to a height of fifteen or twenty feet. They look like bamboos; they are well-nigh impenetrable to a man on horseback; even on foot they make difficult walking unless free use is made of the heavy bush-knife. It is impossible to see through them for more than fifteen or twenty paces, and often for not half that distance. Bears make their lairs in them, and they are the refuge for hunted things."[1]

Canebrakes are thickets of America's only species of native bamboo, a member of the grass family. Cane (*Arundinaria gigantea*), also known as switchcane, grows throughout much of the southeastern United States in a variety of habitats but thrives best in alluvial floodplains. Large historical canebrakes may have developed in fields abandoned by Native Americans decimated by European diseases. Settlers often chose land blanketed in cane because it was high, fertile ground, easy to clear for agriculture, and excellent forage for livestock. A journalist describing Louisiana wrote in 1803: "This reed only grows on land that is never (or almost never) flooded.

. . . These cane brakes, on account of the large amount of humus that they deposit, make the soil very fertile, and the farmers regard their cane brakes as the best possible land; in fact, they judge the quality of the soil by the thickness of the cane."[2] During the Civil War era, cane was more common than in President Roosevelt's day, and the extensive canebrakes on a scale described by early Europeans have disappeared today.

Cane is hollow, light in weight, and when split makes excellent chair bottoms. On April 9, 1865, General Robert E. Lee sat in "a cane-seated armchair beside a square marble-topped table" during the surrender of the Army of Northern Virginia at Appomattox Court House.[3] Soldiers often used it as a building material when available. The swamps and hills in and near Vicksburg, Mississippi, during the 1863 siege provided a particularly abundant source.

Major William J. Bolton, 51st Pennsylvania Volunteers, near Vicksburg, Mississippi, on June 20, 1863: "The bluffs are covered with cane-brakes, blackberry bushes and any amount of underbrush and filled with all kinds of venomous reptiles."[4]

Assistant Commissary John G. Earnest, 79th Tennessee Infantry, at Vicksburg, Mississippi, on Feb. 22, 1863: "From our position, running in a curve to the right was a range of sharp crested hills, shooting up almost perpendicularly to the height of two hundred feet, their sides beautifully shaded with over hanging cane and small trees."[5]

Captain Charles B. Haydon, 2nd Michigan Infantry, near Vicksburg, Mississippi, on June 20, 1863: "Canebrakes such as we buy at home for fish poles are very abundant & are used by the men for almost everything."[6]

Lieutenant John Q. A. Campbell, 5th Iowa Infantry, at Vicksburg, Mississippi, on June 9, 1863: "Today I built myself a bunk with cane, and wrote a letter to the Des Moines Register."[7]

Captain Gabriel Killgore, 17th Louisiana Infantry, at Vicksburg, Mississippi, on June 26, 1863: "A heavy fire was Kept up all night on part of the lines and is going on this morning—8 o'clock a Minnie ball has just struck the cane mattress on which I am lying opposite my shoulder—10 o'clock a Negro Killed in camp."[8]

Sergeant Allen M. Geer, 20th Illinois Volunteers, near Vicksburg, Mississippi, on Nov. 7, 1863: "T. Johnson & I made us a cane bed but slept rather cool."[9]

Major David Pierson, 3rd Louisiana Infantry, in a letter to his father from the Yazoo River above Vicksburg, Mississippi, on Jan. 20, 1863: "Some ingenious Frenchmen have erected houses out of long cane which is to [be] found in great abundance all over the country. You would be amused as well as astonished to see them."[10]

Lieutenant Peter C. Hains, Chief Engineer of 13th U.S. Army Corps, at Vicksburg, Mississippi, on July 30, 1863: "Three sap-rollers were finished to-day. The ground sap-roller was made of solid cane and of the usual dimensions."[11]

Lieutenant Jared Y. Sanders, 26th Louisiana Infantry, at Vicksburg, Mississippi, on June 22, 1863: "Yankees have dug up close to our works all around our lines—so close that they throw over notes & put them on wild canes & hand them to our boys."[12]

Major Samuel H. Lockett, CSA Chief Engineer at Vicksburg, Mississippi: "The causes that led to the capitulation [of Vicksburg] are well known. We had been from the beginning short of ammunition. . . . We were short of provisions, so that our men had been on quarter rations for days before the close of the siege; had eaten mule meat, and rats, and young shoots of cane, with the relish of epicures dining on the finest delicacies of the table."[13]

Lieutenant Robert M. Addison, 23rd Wisconsin Infantry, at Baton Rouge, Louisiana, on June 1, 1864: "Maj Greene ordered the regimental teamsters to haul cane for each company to make shade to protect them from the sun."[14]

William S. Craig, 116th Illinois Infantry, in a letter to his wife from Camp Sherman, Mississippi, on Sept. 8, 1863: "Well, dear, I reckon I have wrote enough for this time so I will close and dream of you again tonight hoping I may dream of you being in a good humor. It is now bedtime and I must crawl on my cane bed and rest."[15]

The dense growth of cane provided cover and hiding places for soldiers and citizens alike.

Kate Stone at Brokenburn Plantation, across the river from Vicksburg near Milliken's Bend, Louisiana, on Dec. 29, 1862: "I am so afraid they [Union soldiers] will get my horse Wonka. . . . Webster has him in charge, hidden in the canebrake."[16]

Brigadier General N. B. Buford, U.S. 7th Corps, at Helena, Arkansas, on May 20, 1864: "The flat county, the narrow roads, the impenetrable thickets of brush and cane afford concealment for guerrillas at every step."[17]

Private Harvey Reid, 22nd Wisconsin Volunteer Infantry, near Brentwood, Tennessee, in February 1863, writing of the recent battle of Spring Hill: "Between 50 and 60 [of the 22nd Wisconsin] took refuge in the cane-brake, . . . but a threat to sweep the brake with canister shot forced their surrender."[18]

Lieutenant John P. Sheffey, 8th Virginia Cavalry, in a letter to his future wife from Greenbrier County, Virginia, on Sept. 16, 1861: "The enemy crowded a cane-patch pretty full about 3 or 400 yards in front of the center, but a few rounds of grape from our central battery slew them and the cane too."[19]

Major John W. Rabb, 2nd Missouri Artillery, at New Madrid, Missouri, on April 10, 1864: "I send you copies of several papers captured by Captain Preuitt, on the dead body of the guerrilla Captain Williams. . . .You will notice that one of the inclosed orders is dated at the Blue Cane. This is a dense canebrake, in the center of which is one of the rebel camps. They here have a store supplied with stolen goods, a distillery, several houses, and a large amount of stock."[20]

CHESTNUT AND CHINQUAPIN

The greatest loss to the ecological integrity of eastern forests since the Civil War has been the loss of American chestnut (*Castanea dentata*) trees.

During the conflict this majestic hardwood species, which towered to 120 feet tall, comprised as much as 50 percent of upland forests from Maine to Alabama.[1] In 1904 a parasitic fungus (chestnut blight) was unwittingly introduced into the United States on imported Chinese chestnuts (*C. mollissima*) and resulted in the decimation of the nonresistant American species.[2] At the time of the Civil War chestnut was the keystone species in eastern forests, producing immense volumes of mast in the form of edible nuts that supported a wide diversity of native birds and mammals and domestic livestock.

Chinquapins (*Castanea* spp.) consist of several species of small trees closely related to chestnuts. Usually less than thirty feet in height, they closely resemble chestnuts and produce similar but smaller edible nuts. Chinquapins suffer from chestnut blight but still manage to thrive in many areas of the east and south.

Utilitarian values of chestnuts and chinquapins during the Civil War were similar. The wood of both species is rot resistant and was desirable for posts and railroad crossties.[3] The larger chestnuts were a prime source of building lumber and also used in the manufacture of boxes, musical instruments, tool handles, barrel staves, and caskets.[4] The bark of both species was a valuable source of tannin for the leather industry. Medicinally, a tea made from the roots was recommended to treat diarrhea in soldiers and as a substitute for quinine. Farmers depended on the fruits to fatten their herds of free-roaming hogs. Humans, too, relished the nuts of both species, parched, boiled, or raw.[5] Chestnuts and chinquapins were the most common wild flora mentioned as food by Confederate soldiers (see table in the chapter "The Civil War Setting").

Jedediah Hotchkiss, topographical engineer of the Army of Northern Virginia, in winter quarters near Fredericksburg, Virginia, on Jan. 3, 1863: "In the evening took the Gen. [Thomas 'Stonewall' Jackson] some apples and chestnuts and then read him the news,—the confirmation of the victory at Murfreesboro."[6]

Major-General W. T. Sherman, USA, in a letter to General Halleck from Summerville, Georgia, on Oct. 19, 1864: "when the rich planters of the Oconee and Savannah see their fences and corn and hogs and sheep vanish before their eyes they will have something more than a mean opinion of the 'Yanks.' Even now our poor mules laugh at the fine corn-fields, and our soldiers riot on chestnuts, sweet potatoes, pigs, chickens, &c."[7]

Private William R. Stilwell, 53rd Georgia Volunteers, in a letter to his wife from Loudon, Tennessee, on Nov. 18, 1863: "I got my pants, shirt and all you sent. You don't know how much good them apples and chestnuts done me. They were so good and then they come from home, from Molly [his wife] and Tommy [his young son]."[8]

Lieutenant Colonel Charles F. Johnson, 81st Pennsylvania Volunteer Infantry, in a letter to his wife from eastern Virginia on Oct. 30, 1861: "I have upon my table a pile of Chestnuts and two cups of Parsimmons brought me by different parties of our scouts, who gathered them while out last night and this morning."[9]

Private Allen M. Geer, 20th Illinois Volunteers, at Jackson, Tennessee, on Oct. 9, 1862: "Went out on a chestnut excursion gathered chestnuts from the trees for the first time."[10]

Private Theodore F. Upson, 100th Indiana Infantry Volunteers, at Kingston, Georgia, on Oct. 9, 1864: "There are the largest chesnut trees here I ever saw, some of them 30 inches across the stump and 75 to 100 feet clear to a limb. The boys have cut lots of them down to get the chesnuts."[11]

Myra Inman at Cleveland, Tennessee, on Oct. 2, 1862: "Pretty day . . . Sister, Mother, Lizzie, Jimmie and I went out chestnut hunting."[12]

Major James A. Connolly, 123rd Illinois Infantry, near Marietta, Georgia, on Oct. 7, 1864: "No enemy encountered until head of column had nearly passed the base of Lost Mountain, which loomed up on right like a huge sugar loaf, its sides covered with scrubby pine, oak and chestnut timber."[13]

Assistant Surgeon Dr. Daniel M. Holt, 121st New York, near Culpeper, Virginia, in a letter to his wife on Oct. 2, 1863: "We had worked hard, as is our custom upon entering a new encampment, in getting our houses in order—setting out shade trees—(pine bushes,) around our tents and policing the grounds in a naturally beautiful grove of chestnuts."[14]

Captain Samuel T. Foster, 24th Texas Cavalry (dismounted), near Atlanta, Georgia, on July 22, 1864: "—woods very thick consisting of Oak Chestnut poplar and undergrowth."[15]

Captain Jacob Ritner, 25th Iowa Infantry, in a letter to his wife from Dallas, Georgia, on June 2, 1864: "There is no house near here where we can get rails or boards. So last night I skinned a big chestnut tree and got a wide piece of bark, lay down on it and spread my gum over me and slept first-rate."[16]

Private Wilbur Fisk, 2nd Vermont Volunteers, near Lewinsville, Virginia, in a letter to his hometown newspaper on Feb. 25, 1862: "this regiment is encamped in the woods, part of which is second growth pine, and the remainder chestnut and oak. The trees outside of the camp have all been cut and used up long ago, and since fuel has become scarce, so scarce that hardly a sound stump can be found in this vicinity which has not been chipped off close to the ground, we have been permitted, in anticipation, I suppose, of warmer and calmer weather, to cut some of the trees inside the camp ground, especially the oak and chestnut."[17]

Lieutenant Charles B. Haydon, 2nd Michigan Infantry, at Hampton Roads, Virginia, on April 16, 1862: "Sergt. Bishop of Co. F was killed while standing near our battery. . . . We wrapped all that could be found of him in his blanket & buried him under an old chestnut."[18]

Assistant Surgeon William Child, 5th New Hampshire Volunteers, in a letter to his wife from near Warrenton, Virginia, on June 24, 1863: "I send you a laurel leaf and flower—a long chestnut leaf and a locust?"[19]

Private Robert D. Patrick, 4th Louisiana, near Ashville, Alabama, on Oct. 23, 1864: "My side is something easier today. Collins and I went up a mountain side and gathered our pockets full of chinkapins. They grow in great quantities here."[20]

Sixteen-year-old Sarah Wadley, daughter of the supervisor of Confederate railroads, near Trenton, Louisiana, on Sept. 27, 1861: "Yesterday we all went chinquapen hunting. . . . We went in the great lumber wagon with four mules . . . first to Mr. Nash's place, where we found 'ever so many' chinquapins . . . we did not gather more than two hours and we had over a bushel of chinquapins when we came to divide them."[21]

Sergeant Edwin H. Fay, Minden [Louisiana] Rangers, near Guntown, Mississippi, in a letter to his wife on Aug. 19, 1862: "Oh my dearest do take care of yourself and not get sick. I know how fond you are of fruit and trash. Please for my sake do not eat it this fall—Chinkapins nor Muscadines. Do not eat such trash for it may give you the same disease our first born died with."[22]

Private Robert A. Moore, 17th Mississippi Regiment, at Culpeper, Virginia, on Aug. 9, 1861: "[We] took a trip to Peidmont Springs. They are about six miles from here. We had a very pleasant trip of it. Saw a great number of chinkapins but they were green."[23]

A. L. Peel, Adjuntant, 19th Mississippi Regiment, near Centreville, Virginia, on Sept. 22, 1861: "Lieut Barksdale myself & some others went out Chinkapin hunting & got a good many."[24]

Captain Samuel T. Foster, 24th Texas Cavalry (dismounted), at Pickett's Mill, Georgia, on May 27, 1864: "Our position is in a heavy timbered section with chinquapin bushes as an undergrowth."[25]

CINCHONA

Some speculate that Alexander the Great died of a stab wound. Others think that the deadly dagger was likely not that of his Persian enemies but rather the proboscis of a malaria-infected mosquito. The relationship between conquest and malaria continued through the ages, destroying armies and civilians alike. Along with destruction of the Inca civilization the Spanish brought malaria to the New World. Ironically, the invasion revealed a secret long known by the natives of Peru—the bark of a certain small tree that grew on steep Andean slopes would relieve fevers, including those of malaria. The natives called it *quinquina,* the "bark of barks." The invaders named the tree Cinchona (*Cinchona* spp.). In 1640 Jesuit priests brought the powerful malaria medicine back to Europe, but it wasn't until 1820 that two French doctors were able to isolate the potent chemical in the bark, now known as quinine. At the time they did not understand that mosquito-borne *Plasmodium* protozoans caused the malarial fevers, and that the medicine worked by suppressing the organisms' ability to multiply. The link between mosquitoes and malaria was not discovered until after the Civil War.[1]

The impacts on the Civil War of malaria and of quinine as an effective treatment are incalculable. Almost a million cases of the disease were re-corded just among Union troops, and entire units were incapacitated at times.[2] In response, the Union's Medical Purveying Bureau alone issued more than one million ounces of quinine compounds, and quinine and its derivative cinchona were the most prescribed medicines for all types of fevers.[3] One author wrote, "Alcohol was the sovereign remedy of the Civil War, rivaled only by quinine."[4] Leaders on both sides realized the value of quinine and sought its procurement in different ways. The only two domestic manufacturers of quinine during the Civil War, Powers and Weightman, and Rosengarten and Sons, were both in the North.[5] Lacking these resources, the South obtained most of its quinine through smuggling operations and the capture of Union supplies. After the war one southern pharmacist wrote:

> The excessive high price of quinine made its handling a profitable employ-ment. Almost every means known to human ingenuity were employed to smuggle it through the lines.... Officers speculating in it, buying and sell-ing until this created a scandal almost equal to that of speculating in cot-ton, and it was finally stopped by a strong proclamation. A large contra-band trade was carried on by an almost continuous line of house-boats floating on the Mississippi river. When opposite Memphis the goods were either sent in at night or into the interior of Arkansas, where trusty par-ties soon disposed of the stock. The great bulk of this trade was sent out by traders and speculators in Paducah, Ky., and Cairo, Ill., and their main points of operation were Memphis, Tenn., Helena, Ark., Napoleon, Ark., and Greenville, Miss.[6]

As the war progressed, the effectiveness of the Union blockade in-creased. The British blockade runner *Nutfield* with a cargo of quinine from Bermuda was run aground by the USS *Sassacus* off the North Carolina coast in February 1864.[7] One southern observer remarked: "Until the lat-ter part of 1863 the supplies of quinine, chloroform, and other medicines were quite sufficient, and only subsequently, when blockade running had become irregular and finally suppressed, did our sick and wounded really suffer for the proper supplies."[8] As a result the South turned more to indig-enous plants as substitutes. These included yellow poplar, dogwood, chin-quapin, willow, and boneset.[9] All treatments were termed "antiperiodics" for the periodic fevers of malaria. As the scarcity of quinine increased dur-ing the war, so did its value, selling at times for $400 to $600 an ounce in the Confederacy.[10] In addition to treating malaria, quinine was first used as a prophylactic in the Civil War, when some northern regiments received a regular dosing of one or two grains per day.[11]

That quinine was considered a valuable treatment for a host of maladies is reflected in the following passage: "The proverbial prescription of the average [Union] army surgeon was quinine, whether for stomach or bowels, headache or toothache, for a cough or for lameness, rheumatism or fever and ague. Quinine was always and everywhere prescribed with a confidence and freedom which left all other medicines far in the rear. Making all due allowances for exaggerations, that drug was unquestionably the popular dose with the doctors."[12]

All Civil War references to quinine relate to its use as a medicament. Generals, privates, and citizens reaped its benefits.

Union Brigadier General Alpheus S. Williams in a letter to his daughter from near Atlanta, Georgia, on July 26, 1864: "My health has been perfect until within a day or so. I begin to feel debilitated and broken down and am taking quinine to build. Shall be all right I hope in a day or so."[13]

Private Robert A. Moore, 17th Mississippi Regiment, in the Confederate hospital at Culpeper, Virginia, on July 30, 1861: "I feel very bad this morning. . . . I took four very large doses of quinine, have had no fever today for the first day in eight."[14]

Kate Stone, Brokenburn Plantation near Milliken's Bend, Louisiana, on Oct. 28, 1861: "Today is but a catalogue of chills. Ashburn and Brother Coley shivered through the morning and burned all the evening. Timely doses of quinine kept them off Sister and Johnny."[15]

Lieutenant Colonel Charles F. Johnson, 81st Pennsylvania Volunteer Infantry, in a letter to his wife from Fair Oaks, Virginia, on June 17, 1862: "The Medical Dept. some time since ordered a half a gill of 'medicinal whiskey' (i.e. whiskey & quinine) to be given each officer and man in the army morning and night."[16]

Private Richard H. Brooks, 51st Georgia Infantry, in a letter to his wife from near Fredericksburg, Virginia, on Aug. 15, 1863: "My Dear I was very sorry to hear of your an the children's sickness but I hope when you get this you will all be better, if you are not try my old remedy take 60 grains of Quinine an forty grains of Rhubarb an put it in one quart of whiskey an take a big spoonful three times a day. an give the children a teaspoon full three times a day. an I think that will cure you all of the chills an fever."[17]

Letter of William L. Nugent, 28th Mississippi Cavalry, in Bolivar County, Mississippi, to his wife on Sept. 28, 1862: "We have in our Company nearly 1/3 of the fighting strength of the Regiment, & have in the last month consumed about two hundred dollars worth of quinine. By the way, did you get the bottle I sent you by Mr. Brown?"[18]

Granville L. Alspaugh, East Feliciana Parish, Louisiana, Skipwith Guards, in a letter to his mother from near Vicksburg, Mississippi, on Nov. 9, 1862: "Ma I am not very well at present I was doing very well before the chills and fever got hold of me but they brought me down a great deal. The boys say that I have some of the hardest chills they ever saw any body have I don't have them regular some times I don't have them [for a] month or two and then again I have them every day for a week. I am taking Quinine now. I took 6 pills Quinine and 1 blue mass pill the Dr said that would break them up he thought."[19]

John B. Jones, clerk to the Confederate secretary of war, in Richmond, Virginia, on Sept. 30, 1863: "His [a Mr. Blair from Mississippi] letter gives a list of prices of medicines in the Confederate States. I select the following: Quinine, per oz., $100; Calomel, $20; blue mass, $20; Opium, $100."[20]

Sergeant Henry C. Lyon, 34th New York Volunteer Infantry, near Harpers Ferry, present-day West Virginia, on March 2, 1862: "Pain in my back very severe. Got up to camping grounds concluded couldn't stand out any longer. Went to the doctor. Took my first dose of Quinine."[21]

British journalist William H. Russell in Washington, D.C., on Aug. 27, 1861: "I was too unwell to do anything for the fever. Took plenty of portwine & quinine by way of cure."[22]

Sarah Wadley, daughter of the supervisor of Confederate railroads, near Trenton, Louisiana, on Sept. 8, 1863: "George has had four chills, but Mother gave him quinine Sunday and he has had none since, he looks very badly and, we fear, is taking the hooping cough."[23]

Sergeant John Q. A. Campbell, 5th Iowa Infantry, in Boonville, Missouri, on Oct. 11, 1861: "The Dr. said this morning that I have the 'intermittent fever' and placed me on the 'quinine list.' I took six quinine pills today."[24]

Reuben A. Pierson, 9th Louisiana Infantry, in a letter to his father from near Centerville, Virginia, on Aug. 25, 1861: "I have had fever for four days but succeeded in breaking it with quinine today. . . . I suppose you will not recognize this as my handwrite. It is not but written by a friend as quinine has made me too nervous to write."[25]

Lieutenant Theodore A. Dodge, 101st New York Volunteers, near Richmond, Virginia, on June 15, 1862: "We draw a ration of whiskey and quinine every day."[26]

Sergeant William D. Dixon, [Savannah] Republican Blues, St. Catherine's Island, Georgia, on Oct. 29, 1861: "I did think at one time that I was going to escape the sickness that has laid the most of the men up, but yesterday I was taken with chill which lasted three hours and a hot fever set in then with a headache which lasted for four hours. . . . I am taking Quinine pills today."[27]

Assistant Surgeon William Child, 5th New Hampshire Volunteers, in a letter to his wife from near Sharpsburg, Maryland, on Oct. 20, 1862: "Last week I had a slight attack of hepatic disease of a periodical nature. I took three doses of 5grs each of quinine in one day. It did me much good."[28]

Private George A. Remley, 22nd Iowa Volunteers, in a letter to his mother from Salem, Missouri, on Christmas Day, 1862: "The Doctor told me the next morning that my headache was beginning to assume a neuralgia form and gave me a few powders, containing about equal portions of quinine, opium and ipecac, which he thought would relieve me."[29]

Private Amos E. Stearns, 25th Massachusetts Volunteer Infantry, at Newport News, Virginia, on Jan. 20, 1864: "I went up to the hospital and got some quinine three times today."[30]

COTTONWOOD

To speak of shoes, it's boots not here;
Our Q.M's [Quartermasters] wise and good,
Give cotton calf-skins twice a year
With soles of cottonwood.
— Hart and Stevens, *The Romance of the Civil War*

Wood from both species of cottonwood (*Populus deltoides* and *P. heterophylla*) found in the eastern United States was used in the manufacture of shoes during the Civil War. Cottonwood thrives in rich alluvial soils of wetlands and is especially abundant along the Mississippi River and its tributaries. Camp Steele, named in honor of Union General Frederick Steele, was located in an area cleared of dense cottonwoods on the east bank of the Mississippi River opposite Helena, Arkansas.[1] An invader with wind-blown seeds, cottonwood is an early succession stage plant that is often the first to colonize sandbars and abandoned riparian fields. Rabbits, beaver, and deer relish the twigs and bark of this fast-growing tree. As steamboat fuel, cottonwood "burned readily when seasoned but did not give a lasting fire, and though widely used for want of something better it was not esteemed."[2] Gunpowder-grade charcoal was made from cottonwood in Alabama and Georgia,[3] and a tea made from cottonwood bark was used to treat chills and fever.[4] Cottonwood lumber is of poor quality and subject to rot but was used when available to build provisional log cabins for winter quarters.

Private Isaac Jackson, 83rd Ohio Volunteer Infantry, at the mouth of the White River in Arkansas, on Nov. 16, 1864: "There are four of us who stay together, we have a regular 'log cabin.' It is about 5 by 10, made of small cottonwood poles and covered with shelter tents."[5]

Corporal Ephraim Anderson, 1st Missouri Brigade, during the siege of Vicksburg, Mississippi, exact date unknown: "A large poplar [cottonwood], over four feet through, which must have been struck from top to bottom by more that two hundred cannon balls, had finally given way and fallen: it was almost entirely severed in a number of places, and shells were buried and still remained in its huge trunk. . . . A party of the Twenty-Seventh Louisiana built up a fire against this tree one morning, and several of them were stooped over and around it, frying their meat for breakfast in the blaze, when a shell buried in the wood was ignited, and exploded in their midst. . . . It was a very remarkable incident, that a shell should explode in the center of a large group of men, and in the very faces of some, without killing any, and only wounding one dangerously."[6]

DOGWOOD

The snowy blooms of flowering dogwoods (*Cornus florida*) are a sure sign of spring from the Atlantic coastal states to east Texas. Several species of dogwoods grow in this area, but none are as well known as flowering dogwood. The actual flowers on this plant are only one-eighth inch in diameter and are surrounded by four large, white, attractive bracts, not petals as many suppose. Flowering dogwood is usually a small understory tree less than thirty feet tall that grows best on well-drained soils.

During the Civil War, a decoction of powdered dogwood bark was used as a substitute for quinine in the South "as it can be easily and abundantly procured" when that valuable drug became scarce. Fevers, especially those associated with malaria, were often treated with dogwood.[1]

The wood of flowering dogwood trees is one of the hardest in North America. One author states that at least ninety percent of the dogwood cut during the nineteenth century was used to make shuttles for the textile industry.[2] Other uses of the dense wood during the Civil War included charcoal, engraving, mallets, tool handles, wedges, plane stock, harrow teeth, hames, horse collars, ox yokes, wheel hubs, barrel hoops, machinery bearings, and cogs in various types of gears.[3]

Sarah Wadley, daughter of the supervisor of Confederate railroads, near Trenton, Louisiana, on April 9, 1864: "we . . . felt quite happy as we drove slowly along in the bright sunshine, admiring at every step some new beauty in the opening springs, now it was the bright green beech trees, now the brilliant white dogwoods or some fragrant clump of honeysuckle and again the glistening green huckleberry leaves or the tiny flowers in the grass, thus happily and quietly we rode along until we came to Trenton and were greeted by the alarming intelligence that the Yankee gunboats were a short distance below Monroe."[4]

Major James A. Connolly, 123rd Illinois Infantry, writing to his wife on April 26, 1864, from camp in Ringgold, Georgia: "The forest trees are putting out leaves and the woods begin to look green; flowers bespangle the green carpet of the valleys, and the white dogwood blossoms are hung in clusters of beauty on the bright green drapery of the mountain side."[5]

British journalist William H. Russell at Wilmington, North Carolina, on April 15, 1861: "White blossoms of dogwood & magnolias beautiful in forests wh. ring with mocking birds, pines everywhere."[6]

Private Richard H. Brooks, 51st Georgia Infantry, in a letter to his wife from Fredericksburg, Virginia, on May 14, 1863: "The Spring is open hear now the Peach trees is all in full bloom an the dog woods are in bloom an all the bushes are buding out nice it makes me think of home."[7]

Southern refugee Kate Stone in Lamar County, Texas, on July 12, 1863: "The prairie we are living on is called a thicket prairie. There are clumps of dwarf dogwood, spice trees, and plums, tangled together with wild grape and other vines and alive with snakes."[8]

ELM

Elms (*Ulmus* spp.) moved the heavy loads of the Civil War. The hard, dense wood was sought because it was "the best wood we have for blocks" in block and tackle rigging used to mount mortars, unload cargo, and hoist sails. Wheel hubs made of elm supported the carriages of cannons and six-mule wagons that carried three thousand pounds of supplies. The half-dozen species of elm in the East were also used in the manufacture of lumber, saddle-trees, ships, tool handles, baskets, chair bottoms, flooring, railroad ties, dye, and woodenware. Rope was made from the inner bark of winged elm (*U. alata*) and used in bagging cotton. Medicine was concocted from slippery elm (*U. rubra*) bark in the form of soothing, jelly-like poultices to treat gunshot wounds and syphilis sores.[1]

> We elms of Malvern Hill
> Remember every thing;
> But sap the twig will fill;
> Wag the world how it will,
> Leaves must be green in spring.
> —Herman Melville

Melville's elms survived the ravages of an epic war, but the continued existence of this species now valued as graceful shade trees depends on another battle, this one with a fungus known as Dutch elm disease.

Felix Poche with CSA Brigadier General Henry Gray's Brigade near Washington, Louisiana, on Sept. 24, 1863: "I spent the whole day at the house, and was quite unwell, still from the bowels. I took some Slippery Elm tea, and it did me some good."[2]

Gideon Lincecum, seventy-year-old naturalist, in a letter to a friend from Washington Co., Texas, on March 26, 1863: "To color drab and light reds, the ley [lye] of some strong ashes is the proper mordant and it is applied after the goods have been boiled sufficiently in the coloring matter. ... The brightest red I have produced on cotton was done with the red elm bark, the wild peach bark next; the goods being washed in ley after thoroughly boiling them in a strong decoction of the bark."[3]

GRAPE

Grapes (*Vitis* spp.) are woody vines that climb with tendrils in search of sunlight. About twenty species of native grapes are found in the eastern United States in a variety of habitats. The well-known fruits of grapes have been consumed by humans for thousands of years and are also important wildlife foods. Fifty-seven species of songbirds have been reported to eat wild grapes as well as bear, coyote, fox, rabbit, raccoon, skunk, squirrel, deer, and catfish.[1] Muscadine grapes (*V. rotundifolia*), also known as scuppernong, grow throughout the Southeast and are popular for their large size and taste in wine and jellies. More than three hundred cultivars of this species have been developed in recent years. Modern research indi-

cates that polyphenols and other nutrients in grapes may have health benefits. During the Civil War, wild grapes were most often mentioned as food,
but the volume of literature pertaining to winemaking using native grapes
prior to the conflict suggests that the age-old art was then a common practice in the United States. One writer of the period lists thirteen varieties of
wines made from native grapes.[2] High- quality inks used to print Confederate currency were often hard to obtain. Wild grapes were crushed and
used as a substitute at times.[3] Soldiers also used grape wood for practical
purposes, such as the construction of sap rollers.

**Private Theodore F. Upson, 100th Indiana Infantry Volunteers, near Sa-
vannah, Georgia, on Dec. 11, 1864:** "We have been making rolling breast
works. We make small rolls of saplings or poles fastened together with
withes or peices of grape vines, and out of these have made a roll about 60
feet long and over 4 feet in diameter. Our sharpshooters get behind this
and roll it down the road, shooting into the Battery embrasures so that
they cannot fire their guns. We have completely silenced two Batteries so
they cannot use them."[4]

**Kate Stone, Brokenburn Plantation near Milliken's Bend, Louisiana,
on Sept. 13, 1861:** "Made John one [a small basket] and he often brings it in
full of muscadines or wild grapes. All of us like wild grape lemonade, especially if feverish. There is generally a pile of the fruit heaped on a side table,
and the boys make and drink lemonade all during the day."[5]

**John S. Jackman, 9th Kentucky Infantry, near Montgomery, Alabama,
on April 30, 1863:** "In the evening went a muskadine hunting on the banks
of the river, and got an abundance."[6]

**Corporal Samuel A. Clear, 116th Pennsylvania Volunteers, at The Wil-
derness, Virginia, on May 5, 1864:** "We pitched into the dark scrubby
pines, grape vines, over old logs, ravines, tree tops, and everything else
that went to make up a wilderness. It was a very irregular line of battle but
it was the best we could do."[7]

Myra Inman at Cleveland, Tennessee, on Sept. 25, 1862: "Cynthia Hard-
wick, Callie and Samantha Swan, John Swan, Peter Smith and I went out
after muscadines this eve."[8]

James C. Bates, 9th Texas Cavalry, near Sherman, Texas, on Oct. 1, 1861: "After dinner took a ride down in Red River bottoms on the hunt of grapes in company with Raz. I found any abundance of small winter grapes, but no large ones. Some of the company bring in grapes of a very large quality—sweet & juicy meet—but a skin so sour that a very few are sufficient to take the skins off our mouths."[9]

British journalist William H. Russell describing the countryside near Mobile, Alabama, on May 19, 1861: "The scuppernong grape native of N. Carolina. The stalk seems to grow rather as a branch than as a vine with a smooth close grained grey bark. Cherokee plums magnolias."[10]

Private Nelson Stauffer, 63rd Illinois Infantry, near Jackson, Tennessee, on Sept. 23, 1862: "On guard at the Humbolt RR Bridge. Plenty of muscadines."[11]

Assistant Surgeon Dr. Daniel M. Holt, 121st New York, in camp near Bakersville, Maryland, in a letter to his wife on Sept. 25, 1862: "Day before yesterday, when we arrived at our present encampment, I got in tired, hungry and almost overcome. Our regiment took line upon the bank of a small stream, on the edges of which grapes—large and luscious hung in great abundance. Parched with thirst and the stomach craving natural acids, induced me to partake (contrary to my knowledge of propriety) freely of them. I have, since that time, had difficulty with my bowels and am to-day almost entirely used up."[12]

Sergeant Hamlin A. Coe, 19th Michigan Volunteer Infantry, north of Atlanta, Georgia, on June 25, 1864: "While it was cool this morning, I went to a swamp and gathered huckleberries enough for a mess and some grapes, and today I have lived upon coffee, hard bread and sauce."[13]

Lieutenant Colonel Charles F. Johnson, 81st Pennsylvania Volunteer Infantry, in a letter to his wife from eastern Virginia on Jan. 16, 1862: "I forwarded, this morning a bundle in care of John Stockten (Sutler), containing my gray jacket & pants and my old blue pants—also a staple from the slave pen at Alexandria, a specimen of petrified wood, and the two pieces of cedar from the grave of [George] Washington and two pieces of wild Grape vine (tied out side) from the *top of the grave*—put them in water Mary and try to cultivate them."[14]

Jedediah Hotchkiss, topographical engineer of the Army of Northern Virginia, near Front Royal, Virginia, on Nov. 3, 1862: "The day was quite raw. We enjoyed the profusion of delicate, wild grapes that grew by the road."[15]

Lieutenant John P. Sheffey, 8th Virginia Cavalry, in a letter to his future wife from Fayette County, Virginia, on Oct. 30, 1861: "Like Canaan this is a land of milk and honey—and buckwheat cakes and grapes, not grapes indeed of Eshcol, but of a more deliciously flavored species, known in the elegant phraseology of the country as 'Possum Grapes.'"[16]

  ❧ Grape Shot—cannon projectile similar to canister consisting of iron or lead balls trussed up with fabric and twine (and thus resembling clusters of grapes) or held together with iron rings; used most often as naval ammunition.[17]

  ❧ *Scuppernong*—Confederate side-wheel steamboat destroyed by fire set by Union sailors in June 1862 in North Carolina.[18]

HERBS

Herbs can be broadly defined as nonwoody plants (usually perennials) that humans use for culinary or medicinal purposes. For thousands of years until well into the twentieth century the use of herbs to treat various ailments was standard procedure, and herbs as flavoring for foods are as popular today as during the Civil War. Herbs are found in a wide range of taxonomic families and grow in a variety of habitats. Many herbs, especially mints, were brought to America by European settlers and later became naturalized. Settlers also quickly learned from Native Americans the benefits of native herbs and adopted their uses. Following are examples of herbs used during the Civil War.

Peppermint (*Mentha* spp.) usually refers to a European species, but similar native mints such as mountain mint (*Pycnanthemum* spp.) and horsemint (*Monarda* spp.) are common, and all were used to treat intestinal disorders.[1]

Sarah Wadley, daughter of the supervisor of Confederate railroads, as a refugee traveling in the Tensas Swamp of northeastern Louisiana on Sept. 30, 1863: "I was awakened by a dull pain in my side which increased to agony, it gave me not a minutes rest for two hours, I suppose, when it lulled and I dozed a little while, then it came on worse than before and I threw up the supper I had eaten, this waked Father and then Mother, they did everything they could think of, I took peppermint, laudanum, and number six, all without any effect on the terrible pain, the worst I have ever known."[2]

Harriett Goodwin Pierce, Union soldier's wife, in Westford, Vermont, on June 5, 1864: "Not a very pleasant day. We went up on the Hill to get wintergreen."[3]

Refugee Ellen Renshaw House in Eatonton, Georgia, on Jan. 29, 1865: "This afternoon . . . I walked out to old Mrs Pikes to get Mr Pleas some Catnip."[4]

Lieutenant Richard Goldwaite, 99th New York Volunteers, in a letter to his wife from Fort Monroe, Virginia, on Aug. 30, 1862: "I must look out for myself and take a little more Horse Mint Tea as I feel if I am going to have a shake tonight."[5]

By the time of the Civil War, dandelion (*Taraxacum* spp.) was a well-established and frequently used herbaceous weed from Europe. Various parts of the plant were sought to treat a gang of ailments including itch, gallstones, and jaundice. The ground, roasted roots also served as a coffee substitute.[6]

Lieutenant Theodore A. Dodge, 101st New York Volunteers, near Harrison's Landing, Virginia, on July 16, 1862: "I must get some medicine in lieu of dandelions, of which we can get none here. I think some good liquor would do me good."[7]

Harriett Goodwin Pierce, Union soldier's wife, in Bradford, New Hampshire, on May 7, 1865: "This has been an afful lonely day and qite cold. I slept away a goodeal of the day. Dug some Dandelions."[8]

Teas thought to have therapeutic value were made from the fruit (hips) of wild roses (*Rosa* spp.). Their only likely value was as a source of Vitamin C, unknown to people of the Civil War period.[9]

Private Nelson Stauffer, 63rd Illinois Infantry, near Bridgeport, Alabama, on Dec. 4, 1863: "we scattered through a wide Prairie to gather the berrys of the wild rose bush."[10]

Black cohosh (*Actaea racemosa*), also called black snakeroot or bugbane because the flowers repel insects, is a native forest herb of the eastern United States. American Indians introduced the practice of using a brew from the boiled rhizome to treat menstrual cramps, rheumatism, and sore throat.[11]

Corporal George M. Englis, 89th New York Volunteers, in a letter to his mother from Folley Island, South Carolina, on Oct. 14, 1863: "The medicine you sent for the sore throat is very good, although I have got lots of [black] cohosh that came from home and some I dug in Maryland. I never have had the sore throat since I have been south."[12]

Seneca snakeroot (*Polygala senega*) was also known as rattlesnake root because Seneca Indians considered it an antidote for rattlesnake bites. Doses of the root extract were considered effective treatment of many respiratory ailments including asthma, pneumonia, and bronchitis, in addition to chronic rheumatism.[13]

Major James C. Bates, 9th Texas Cavalry, near Tuscaloosa, Alabama, in a letter to his mother on May 4, 1864: "Sister: I have opened this to give you a remedy for headache. It has frequently relieved me from a severe headache in half an hour. It is simply, equal parts of cloves & Seneca snake root, finely pulverized, & used as snuff. Don't laugh until you try it."[14]

Goldenclub (*Orontium aquaticum*) is an early blooming, aquatic plant of swamps and marshes with broad blue-green leaves that repel water, thus the colloquial name "never wets."[15]

Confederate sympathizer Myra Inman at Cleveland, Tennessee, on April 10, 1863: "Mother, Johnnie, myself and Sues went out to Mrs. Traynor's big spring to get some 'never wet lillies' to put on Sues' burnt leg."[16]

Lobelia is the genus and common name of several woodland plants with attractive blue or red flowers considered during the Civil War era as "one of the most valuable of our indigenous plants." Once used to treat syphilis, asthma, bronchitis, tuberculosis, and nervous disorders, the actual results were likely negative as the plants contain highly toxic forms of alkaloids, similar to nicotine in physiological effects.[17]

Emily B. Moxley in a letter to her husband in the 18th Alabama Infantry from Pike Co., Alabama, on Jan. 13, 1862: "Pa thinks now that Dr. Dyer killed Brother. He gave him Lobelia on Sunday before he died and kept him throwing up all day and such strain[in]g you never saw."[18]

Arrowroot is the common name of several species of perennial plants with starchy rhizomes. *Maranta arundinaceae,* a native of rain forests that was naturalized in south Florida by the time of the Civil War, was cultivated or harvested from the wild and laboriously processed to obtain arrowroot flour. Highly valued as a food for sick and convalescing patients, arrowroot from the West Indies was often cargo for southern blockade runners. The *Colonel Long,* a Confederate fishing schooner, was loaded with bags of arrowroot when captured by the USS *Jamestown* off the Georgia coast in September 1861, and the British sloop *Lizzie* contained arrowroot when overtaken by a Union vessel in August 1862 off the North Carolina coast.[19]

Confederate nurse Kate Cumming at a hospital in Corinth, Mississippi, on April 30, 1862: "We have a quantity of arrow-root, and I was told that it was useless to prepare it, as the men [patients] would not touch it. I thought that I would try them, and now use gallons of it daily. I make it quite thin, and sometimes beat up a few eggs and stir in while hot; then season with preserves of any kind—those that are a little acid are the best—and let stand until it becomes cold. This makes a very pleasant and nourishing drink; it is good in quite a number of diseases; will ease a cough; and is especially beneficial in cases of pneumonia."[20]

Union nurse Hannah Ropes at a hospital in Washington, D.C., in a letter to her daughter in November 1862: "I sent up [to a patient] a cup of arrowroot seasoned with wine.... Last night I went again to see him. He certainly is better. Who knows but this Virginian may live to tell of the war to his grandchildren?"[21]

Georgeanna W. Bacon, Union nurse, in Washington, D.C., on March 12, 1862: "While we were cooking some arrowroot in our parlor for a Vermont private, sick in this hotel, Joe came in."[22]

Onions, garlic, shallots, and leeks (*Allium* spp.) are all members of the lily family that can be considered both medicinal and culinary herbs. Many species are native to America and other favorites were brought with settlers from Europe. As a group they are the most ancient and popular herbs in the world. By the time of the Civil War members of this genus had been used to treat most known ailments in some form or fashion. They were valued as aphrodisiacs and to treat intestinal parasites, earache, snakebite, colds, high and low blood pressure, and to ward off a host of evil spirits. Onions and garlic, along with bell peppers, compose the "Holy Trinity" of spices used in many southern dishes.[23]

General Robert E. Lee from Headquarters, Army of Northern Virginia, on March 27, 1863: "The men are cheerful, and I receive but few complaints; still, I do not think it is enough to continue them in health and vigor, and I fear they will be unable to endure the hardships of the approaching campaign. Symptoms of scurvy are appearing among them, and to supply the place of vegetables each regiment is directed to send a daily detail to gather sassafras buds, wild onions, garlic, lamb's quarter, and poke sprouts, but for so large an army the supply obtained is very small."[24]

Frederic A. James, U.S. Navy, while serving as a prisoner of war at Salisbury, North Carolina, on May 14, 1864: "We have also been able to buy a few leeks, or wild onion tops, which when cut up & cooked in our soup, are very good. Some of our fellows have 'gone to grass,' & eat boiled clover for greens."[25]

A. L. Peel, Adjutant, 19th Mississippi Regiment, near Fredericksburg, Virginia, on May 8, 1863: "I walked about over fields today looking for my horse, could not find him. Gathered some wild onions for dinner."[26]

Private William R. Stilwell, 53rd Georgia Volunteers, in a letter to his wife from near Fredericksburg, Virginia, on April 21, 1863: "I still have to complain of not much to eat. Now that spring has come I want some vegetables and can get none except the wild shallots which grow here in abundance. We fry them and they eat finely. Southern people were never made to starve. We can lie in the woods and whip Yankees."[27]

HUCKLEBERRY

The term *huckleberry* is a catchall used to identify about two dozen species of similar shrubby plants known for their edible fruits. Huckleberries (*Vaccinium* spp. and *Gaylussacia* spp.) are usually differentiated from blueberries by having more seeds, although laymen of the Civil War era didn't bother with the distinction. The numerous species are adapted to a variety of habitats, with many preferring acidic soils. A tea made from the bark, roots, and leaves of one type was used to treat diarrhea and "sore mouth." The bark was used for tanning leather.[1] Huckleberries are important wildlife foods, and foraging soldiers competed with songbirds, bears, and mice for the nutritious fruits. Second in importance only to blackberries as a source of wild fruit, huckleberries were gathered at every opportunity by soldiers of both sides.

Private James H. Avery, 5th Michigan Cavalry, near Smithsburg, Virginia, on July 24, 1863: "We were in a field of brush and timber in which grew plenty of huckleberries, nice and ripe, and as we advanced, we would pick berries and then fire at the rebs, spite of the shells, which came pretty thick, we were not going past our berries without eating a share."[2]

Private William R. Stilwell, 53rd Georgia Volunteers, in a letter to his wife from near Richmond, Virginia, on July 22, 1862: "I have a great deal of playtime. I can go anywhere when I am not on guard. The other day my messmate and I, for I have but one, went huckleberry hunting. We had the good luck to get a good many. We had some sugar that we drew and we concluded we would have some pies. So we set to work and made some five or six and they were splendid too, just such as are sold to the boys for fifty cents."[3]

Private Nelson Stauffer, 63rd Illinois Infantry, near Kingston, Georgia, on July 11, 1864: "Went huckleberrying. On fatigue loading corn sacks."[4]

Sergeant Alexander Chisholm, 116th Pennsylvania Volunteers, in a letter to his father from near Petersburg, Virginia, on Aug. 30, 1864: "Huckleberries were ripe and plenty, I ate my fill as we passed through the woods."[5]

Sergeant John Q. A. Campbell, 5th Iowa Infantry, near Boonville, Missouri, on June 4, 1862: "I gathered a cup of huckleberries and had a good stew for dinner."[6]

Private Wilbur Fisk, 2nd Vermont Volunteers, near Richmond, Virginia, in a letter to his hometown newspaper on May 20, 1862: "To-night we are encamped in an oak woods, whose rich foliage protects us overhead while huckle and blue berries just in full bloom make a beautiful carpet underneath."[7]

JUNIPER

The tree referred to most often as "cedar" during the Civil War was not a cedar at all but rather a juniper, confusingly known today as eastern red-cedar (*Juniperus virginiana*). An evergreen with fragrant needles and small cones, eastern redcedar is found throughout the eastern United States. This species prefers limestone regions but is common on many poorer soils.

The heartwood is aromatic, very resistant to rot, and repels insects, making it desirable for shingles, posts, barrel staves, and boat building. Easy to work, soldiers used it for buckets, and Confederates carried cedar canteens on the Gettysburg battlefield.[1]

As a medicinal, various decoctions of "cedar" shavings, berries, and leaves were used to treat rheumatic pains, joint swelling, and blisters. Even a fungus commonly found on the tree was employed in the eradication of internal parasites.[2]

Eastern redcedar is a distinctive, easily recognized tree. As such, the term *cedar* was commonly used in the description of surroundings, especially in Tennessee.

John S. Jackman, 9th Kentucky Infantry, during the battle of Murfreesboro or Stone River, Tennessee, on Dec. 30, 1862: "In the afternoon, the rain having slacked up, I sought out the regiment, which I found lying about in a cedar thicket, grumbling about the weather. Sometimes, too, shells would come tearing through the cedars, making a fellow feel uncomfortable."[3]

Lieutenant Henry W. Reddick, 1st Florida Infantry, writing of the December 1864 retreat from the battle of Franklin, Tennessee: "we went through the old cemetery [near Murfreesboro] where our boys were buried that had fallen in the first fight there about two years before. As we fell back through the cedar grove I noticed a number of places where the bones of those buried there in the fight before were sticking up through the ground."[4]

Lieutenant John Q. A. Campbell, 5th Iowa Infantry, near Cowan, Tennessee, on Nov. 13, 1863: "Timber on the mountain—hickory, oak, ash, sugar, chestnut, etc. on the side of the mountain among the rocks, where there is good soil—red cedar."[5]

Private William G. Bentley, 104th Ohio Volunteer Infantry, in a letter to his family from near Richmond, Kentucky, on Dec. 12, 1862: "The opposite bank [of the Kentucky River] must be 300 or 400 feet high in some places, of solid limestone, smooth and white with little cedar bushes growing out of the crevices."[6]

Private Robert M. Holmes, 24th Mississippi Volunteers, near Eagleville, Tennessee, on Feb. 16, 1863: "Very soon it began to rain slowly but study so it was very bad marching in the mud & water. . . . We then put up our tents in the cedar thicket while the rain was falling & prepared for the nights rest the best we could."[7]

During the long, boring months when armies were in winter quarters, efforts were often made to enhance the aesthetic landscape of the camps by laying out streets, organizing neat rows of cabins or tents, and even planting trees. Occasionally, they were also planted for shade.

Major James A. Connolly, 123rd Illinois Infantry, Murfreesboro, Tennessee, on May 1, 1863: "We have cedar trees about ten feet high set out in all the streets through our camp and a double row of them about 12 feet high set out the whole length of our headquarter street, which makes a fine shady avenue for these warm days. I have my table out under one of these cedars while writing this letter, and in the next tree to me, right in front of my tent, a couple of red birds have their nest and greet me with fine music every morning about daylight."[8]

Cedar, because of its durability and fragrance, was often used in the construction of huts when available. Bedding made from the boughs was pleasantly aromatic in a shack or hospital crowded with unwashed soldiers.

Private Wilbur Fisk, 2nd Vermont Volunteers, near Kernstown, Virginia, in a letter to his hometown newspaper on Nov. 16, 1864: "It [a soldier's hut built for winter quarters] was built of large, straight cedar sticks, from three to five inches through, hewn smooth on the inside, and papered with newspapers. He (or they) had a board floor, brick hearth, glass window, and as good a chimney for its purpose as ever was built."[9]

Private William G. Bentley, 104th Ohio Volunteer Infantry, in a letter to his family from Lancaster, Kentucky, on April 3, 1863: "We built huts out of cedar boughs and were just ready to turn in for the night when orders came for Co.'s G & B to go up to the train as guards."[10]

Sergeant Charles B. Haydon, 2nd Michigan Infantry, near Washington, D.C., on June 28, 1861: "I find a good deal of trouble in keeping the ground dry & good in the tent where I sleep. If you use straw it soon gets damp & mouldy & full of bugs & crickets & all manner of vermin. I afterwards tried cedar boughs. These keep the insects out but the ground gets damp & mouldy under them."[11]

Confederate nurse Kate Cumming in Griffin, Georgia, on April 18, 1865: "But there is a great lack of shade trees; I tried the other day to get some cedar [boughs] to dress our wards, but failed. How often I think of the grand old woods of Tennessee and North Georgia."[12]

Eastern redcedar was used opportunistically for firewood. As Christmas trees and seasonal garlands, such use continues today.

Capt. Samuel T. Foster, 24th Texas Cavalry (dismounted), near Nashville, Tennessee, on Dec. 11, 1864: "Still froze up, all quiet on the picket line. I have dug out a fire place in the side of the ditch, and burn cedar rails."[13]

John B. Jones, clerk to the Confederate secretary of war, in Richmond, Virginia, on Dec. 25, 1863: "It is a sad Christmas; cold, and threatening snow. My two youngest children however, have decked the parlor with evergreens, crosses, stars, etc. They have a cedar Christmas-tree, but it is not burdened. Candy is held at $8 per pound."[14]

LOCUST

Three different tree species were referred to as "locust" during the Civil War. Each has thorns and compound leaves and produces beanlike seed pods. Honey locust (*Gleditsia tricanthos*) and water locust (*G. aquatica*) are closely related and armed with formidable, branched three-inch thorns. Honey locust grows throughout the eastern United States, while water locust is usually confined to bottomland hardwood swamps. Black locust (*Robina pseudoacacia*), with short, unbranched thorns, was once more restricted in range but has become naturalized in many areas as a result of introductions for utilitarian purposes. The seed pods of all three species are of moderate value as wildlife food.

Although all three species were renowned for the durability of their wood and sought for fence posts, railings, crossties, and bridge timbers, most Civil War references to "locust" pertain to black locust, which lasts fifty years or more in contact with the soil.[1] The stiffness of black locust wood was valued in shipbuilding. In referring to naval architecture one writer of the era stated: "In many places where strength is wanting, locust timber will bear a strain which would break oak of the same size. Thus an oak tiller has been known to break near the head of the rudder in a gale of wind, which has never happened with a locust one."[2] Black locust was used for tool handles, cart shafts, and wooden platters and spoons.[3] As firewood the fuel value of dried black locust was equal to coal.[4]

Honey locust wood was used in axletrees for carriages and wheel hubs. The fermented pods produced a beer. This species was also called "Confederate Pintree" because the long thorns were used to pin together the ragged uniforms of southern soldiers during the later years of the war.[5]

Union sympathizer Frances Peter in Lexington, Kentucky, on Sept. 13, 1863: "I have been looking from Ma's window at the soldiers in the lot. How prettily they are grouped, some standing some sitting around their camp fire, where their evening meal is cooking. If I were but artist enough what a nice sketch in colors it would make! The soldiers in their blue uniforms, surrounded by the white tents, the blazing fire with its column of blue smoke rising up amid the tall black stemmed locust trees, against some of which a shining rifle or two is leaning, and the carpet of 'Blue' grass looking so fresh and green after the rain and contrasting with the bare brown space around the fire."[6]

Confederate sympathizer Myra Inman at Cleveland, Tennessee, on Jan. 15, 1863: "We can only hear the lonesome low moan and wail of the wind as it sighs through the bare branches of the catalpas and locust, reminding us that all is not sunshine in life."[7]

Lieutenant Henry W. Reddick, 1st Florida Infantry, writing of the Nov. 30, 1864, battle of Franklin, Tennessee: "When they [Union soldiers] fell back to their main line our boys began dropping like corn before a hail storm, and we never did succeed in reaching their main line, for about fifty yards in front of it they had cut down a lot of thorny locust bushes and it was impossible in face of the hot fire to get through them."[8]

Captain Jacob Ritner, 25th Iowa Infantry, in a letter to his wife from Young's Point, Louisiana, on March 30, 1863: "We have had some warm weather here—the trees are all green with leaves. One of the men brought me a nice nosegay of locust and other blossoms yesterday."[9]

🌿 An ancient honey locust tree in Gettysburg National Military Park that had stood witness to the pivotal battle was severely damaged by a strong thunderstorm on August 7, 2008.[10]

MAGNOLIA

Perhaps no tree is more emblematic of the Deep South than the southern magnolia (*Magnolia grandiflora*). Large, lustrous evergreen leaves and fragrant white flowers nine inches in diameter were cherished as ornamentals by southerners and admired by invading northerners. Nine species of the magnolia family are found in North America, all in the eastern United States. Eight are in the *Magnolia* genus with the other, *Liriodendron* (*L. tulipifera*, yellow-poplar), being the only one with significant commercial value as lumber.

Confederate surgeon Francis Porcher reported that various parts of trees in this family were used as a laxative, and to treat fevers, headaches, and worms. Horses with stomach disorders or infected with bots were alleged to benefit when dosed with decoctions of yellow-poplar bark. The presence of sweetbay magnolia (*M. virginiana*), which grows on wet sites, was thought to prevent such areas from "generating malaria."[1] This may have led to the belief that the bark of some magnolias was a favorable substitute for quinine.

As the tallest North American flowering plant, yellow-poplar with its straight trunk was sought for building logs in log structures of all types. It was also used for lumber, furniture, cabinetry, and carriage panels. Wood from other species in the family was used for pump logs and troughs, baskets, and woodenware.[2]

James T. Ayers, Union recruiter, near Huntsville, Alabama, on Sept. 10, 1864: "And then here you find the magnolia tree planted by nature herself as tho oald nature was Determined to Crown with Beauty all its other works in Beauty with this Beautiful tree. Here she is in all her native glory with her Broad green Leaves big white Roses Look when in full Bloom Like A great white monument handsomely doted with those big green Leaves. Beautiful is the Magnolia tree and those Large fine flowers are of A most Delishious fragrance and Smell."[3]

Major James A. Connolly, 123rd Illinois Infantry, near Robertsville, South Carolina, Feb. 6, 1865, in a letter to his wife: "The soil is very poor and sandy here. Magnolia and cypress trees grow very large. Our men chop down splendid magnolias to make bridges of and to corduroy roads. As the magnolia timber is not found in the North, I got off my horse yesterday as we were marching along, and picked a chip off the end of a log about two feet in diameter, which I enclose."[4]

Eliza W. Howland, Union nurse, in a letter to her husband from a hospital ship near White House, Virginia, in June 1862: "There are large bunches of laurel and magnolia in our parlor-cabin and dining room, and the air is full of their fragrance."[5]

Private Theodore F. Upson, 100th Indiana Infantry Volunteers, near Savannah, Georgia, in December 1864: "This is a beautiful City and very old. . . . The famous Shell Road is more than 5 miles long and smooth as a house floor with great magnolia trees on either side."[6]

Sergeant Taylor Peirce, 22nd Iowa Infantry, writing to his wife on Jan. 29, 1865, at Savannah, Georgia: "The wood has a very diversified appearance. Some of the trees having shed their leaves while the evergreens such as the large magnolia live Oak and Pine growing up in the midst gives a very pleasing affect to the view."[7]

🌿 Magnolia Hall was the last large antebellum home built in Natchez and is one of the best examples of Greek Revival style still standing. Shells from the Union gunboat *Essex* damaged the house during the war.[8]

MAPLE

"Reveille," "Taps," "The Long Roll," "The Rogue's March"—these drum calls brought forth emotions as varied as the men they summoned, both Union and Confederate. All reverberated from drums made of hard maple (*Acer* spp.). Fifes too were turned of maple but were thought to be inferior to those built of exotic hardwoods.[1]

The nine species of maples found in the eastern United States are small to medium-sized trees with opposite leaves. Often divided into "hard maples" that yield excellent lumber and the more brittle "soft maples," they grow in a variety of habitats from low swamps to mountaintops. Their winged seeds, buds, and flowers are eaten by a host of birds and small mammals.

During the Civil War, sugar maple (*A. saccharum*) was the most important commercial species in this group both for the lumber it produced and the sugar made from sap collected in the spring by tapping trees. Maple sugar was the most common sweetener in the country, being more available and less expensive than cane sugar from the Deep South. In the years leading up to and during the war, some northerners considered it unpatriotic to use cane sugar produced with slave labor. As sugar maple grew primarily in northern states, Union soldiers benefited most from a food source that did not spoil during shipment and storage.[2] In addition to musical instruments, maple wood was used in gunstocks, saddle trees, shoe lasts, woodenware, shipbuilding, farm tools, furniture, flooring, charcoal production, and as fuel.[3] One Civil War doctor reported that a decoction of red maple (*A. rubrum*) bark administered as a wash improved eye disorders.[4]

General Robert E. Lee in a letter to his daughters from Valley Mountain, Virginia, on Aug. 29, 1861: "The mountains are magnificent. The sugar-maples are beginning to turn already, and the grass is luxuriant."[5]

Captain Theodore A. Dodge, 101st New York Volunteers, near Stafford, Virginia, on Jan. 15, 1863: "The orders say the teamsters are to cut leaves and limbs of the maple, elm, and other soft trees, which furnish a good substitute for hay. Where I wonder are we to get these in a section of country which only produces fir and pine. Rather a precarious footing our poor horses will stand on."[6]

Private Isaac Jackson, 83rd Ohio Volunteer Infantry, near Paris, Kentucky, on Oct. 21, 1862: "The scenery is very beautiful. . . . There are plenty of sugar [maples] & Hickory trees here. Our camp was in a nice sugar grove, and on one side of us at the foot of the hill is another nice sugar grove that looks very much like Godley's Grove used to."[7]

🍁 *Maple Leaf*—a Union troop transport ship. In 1863 a group of Confederate prisoners of war on the vessel escaped after overpowering their guards and taking control of the ship.[8]

MISTLETOE

To the Druids, mistletoe (*Phoradendron* spp. et al.) appeared to spring from thin air. Equally strange, it seemed to defy nature by living its entire life high in the branches of trees, never descending to earth, a plant's natural habitat. For these reasons they declared mistletoe and the oak trees on which it grew to be sacred. More than twenty species of mistletoe grow in North America, with most of those in the Civil War arena living on broad-leaved hardwood trees. Mistletoe is parasitic on its host tree, deriving most of its water and nutrients in the form of minerals from the branches to which it is attached. Although a heavy growth of mistletoe may contribute to the decline of a tree with other ailments, it doesn't usually kill its host. Fruitivorous songbirds such as bluebirds and cedar waxwings relish the white berries that form in late autumn. Mistletoe was once used to treat "fits" of epilepsy but is now considered toxic to humans.[1] The tradition of kissing under a mistletoe sprig during the holidays had apparently evolved by the Civil War.

Dr. Francis Bacon, Union physician, in a letter to his wife from Tybee Island, South Carolina, on Dec. 24, 1861: "We have all the foliage orthodox for Christmas here, including holly and mistletoe with berries of scarlet and white wax. The jungly unscarred forest of this island is superb."[2]

Kate Stone, Brokenburn Plantation near Milliken's Bend, Louisiana, on Dec. 29, 1862, writing of her preparations for Christmas: "Johnny and I gathered a lot of mistletoe and crimson casino berries, and we decorated the parlor and hall prettily next day."[3]

Sergeant Lycurgus Remley, 22nd Iowa Volunteers, in a letter to his father from Milliken's Bend, Louisiana, on April 11, 1863: "This flowerless, black looking specimen of vegetation is a cutting from a mistletoe, a parasite very common down here."[4]

John M. Follett, 33rd Illinois Volunteer Infantry, at Pitman's Ferry, Arkansas, in a letter to his wife on April 25, 1862: "You spoke in one of your letters about my sending you a sprig of mistletoe. I wrote you that I did not know what it was, but there is lots of it here. I will send you some."[5]

Colonel Thomas W. Higginson, 1st South Carolina [African American] Volunteers, near Beaufort, South Carolina, on Feb. 23, 1863: "Spring advances, grass grows green, yellow and fragrant jasmines flaunt from tree tops in the woods, above white waxen beads of mistletoe, which I have never seen before."[6]

🖛 USS *Mistletoe*—Fifty-ton steam tug used by the Union Navy on the Western Rivers.[7]

MULBERRY

The Civil War may be indirectly implicated in the continuing devastation of American forests by an army of insects proven to be less stoppable than the armies of the Union or Confederacy. When southern cotton became unavailable in the North, Leopold Trouvelot, a Boston naturalist, accelerated his research into producing a viable silkworm for the northern textile industry. It led to his infamous, late 1860s importation of gypsy moth eggs from France in an effort to crossbreed them with silkworm moths. Because the two insects were only distantly related, they could not interbreed and the experiment failed. Either accidentally or intentionally, gypsy moth caterpillars were soon released into a hospitable environment and became the gypsy moth plague that causes damages valued in the hundreds of millions of dollars to North American forests each year.[1]

The intended host plants for the silkworm caterpillars were native red mulberry (*Morus rubra*) and white mulberry (*M. alba*), an introduced Asian species that serves as the foundation of silk production in other parts of the world. Promoted by the U.S. government, white mulberry was widely planted before the Civil War in an unsuccessful effort to develop a domestic silk industry. Like the gypsy moth, white mulberry has become an invasive species with negative ecological impacts on natural areas. It hybridizes with and transmits disease harmful to red mulberry and displaces other native vegetation.[2]

Red mulberries are small- to medium-sized trees that grow throughout the eastern United States. Many species of birds and small mammals relish their raspberrylike fruits. Choctaw and Natchez Indians made cloth from the bark. Settlers made rope, cordage, brown paper, and chair bottoms. The wood is very durable and during the war was used for fence posts, cooperage, crossties, and boat construction. One writer of the period stated: "it [red mulberry wood] is employed in naval architecture at Philadelphia and Baltimore, for the upper and lower parts of the frame, for knees and floor timbers, and for tree-nails."[3] A Confederate surgeon reported that the fruit was used as a laxative and as a refreshing drink.[4] Soldiers and civilians occasionally wrote of eating the fruits when they found them.

Major James A. Connolly, 123rd Illinois Infantry, writing to his wife on June 9, 1864, from camp near Acworth, Georgia: "I am sitting in the door yard of a 'Georgia planter,' under the shade of his mulberry trees, the ripe fruit hanging above me. Think I shall climb the tree and eat some of it after finishing this."[5]

Kate Stone, Brokenburn Plantation near Milliken's Bend, Louisiana, on May 27, 1861: "Ashburn and Johnny, the youngest of the boys, brought us some mulberries from their ride in the woods, but nobody but children cares to eat mulberries."[6]

Assistant Surgeon Dr. Daniel M. Holt, 121st New York, near Charles City, Virginia, in a letter to his wife on June 15, 1864: "The mulberry, too, is a fine tree, from which I have frequent repasts of the blackberry-like fruit which it yields."[7]

Private William G. Bentley, 104th Ohio Volunteer Infantry, in a letter to his family from Mount Vernon, Kentucky, on June 14, 1863: "I was foraging this morning and got some light biscuits and all the mulberries I could eat."[8]

John Hay, assistant secretary to President Lincoln, near Beaufort, South Carolina, on May 23, 1863: "Bannister having a convalescent appetite went off with me to eat plums & mulberries."[9]

Private S. O. Bereman, 4th Iowa Cavalry, near Vicksburg, Mississippi, on May 19, 1863: "Went on to within 4 miles of Vicksburg and went into camp. We were told it was much healthier out here than closer to the city! Got all the mulberrys we could eat, but it dont last long."[10]

OAKS

Timbers hewn from mighty oaks (*Quercus* spp.) framed the ships of both Union and Confederate navies. More than thirty species of oaks are native to the eastern United States. Hybrids among these are common. Oaks are often divided into two major groups of white oaks and red oaks. The white oak group has leaves with rounded edges, and their acorns mature in a single season. Examples include white oak (*Q. alba*), post oak (*Q. stellata*), chestnut oak (*Q. prinus*), and chinkapin oak (*Q. muehlenbergii*). The leaves of red oaks are bristle-tipped, and their acorns take two years to mature. Examples include northern red oak (*Q. rubra*), black oak (*Q. velutina*), pin oak (*Q. palustris*), and water oak (*Q. nigra*). Oaks were found in all of the major hardwood ecosystems in the Civil War arena. In the period since the conflict oaks have declined significantly. Industrial forestry has converted oak habitats to pine monocultures, and more than seven million acres of oak forests were cleared for agriculture in the Mississippi River Valley. Oaks were and continue to be of major importance to wildlife. Acorns provide an abundant staple, especially during the winter, for a host of large and small mammals, waterfowl, and songbirds. Oak leaves and twigs are used as nesting materials, and hollow trunks provide dens. During the Civil War humans too used oaks—and for much more than shipbuilding.

As a group the whites oaks have tight-grained wood less porous than that of red oaks. White oaks were thus sought for cooperage and shingles, and for components requiring great strength such as tool handles and the hubs, axles, and tongues of wagons.[1] White oak splits were woven into chair bottoms and baskets of various dimensions, from gizzard-shaped egg baskets to the giant plantation cotton baskets. They were used as binding for cotton bales.[2] Red oaks contain high concentrations of tannic acid, vital in the Civil War tanning industry. In 1860 one southern magazine writer urged, "Every farmer ought to save all the tan-bark that he can, for we speak advisedly when we say that the Confederate States are even now short of oak bark if they are to manufacture all the leather they are to consume in saddles, bridles, harness, saddle-bags, buggy and carriage harness, caps and hat linings, book bindings, boots and shoes."[3] White oaks have less tannic acid, making their acorns more palatable for wildlife, livestock (especially hogs), and humans. Undoubtedly, hungry Civil War soldiers continued the traditions of Indians and colonists at times by brewing acorn coffee and eating acorns prepared in various ways, from raw to roasted to boiled mush. Extracts from red oak bark were used to make yellow, green, and black dyes for wool and silk. In 1863 a Texas writer penned, "For [dyeing] wool: let the wool be boiled for an hour, or longer, with about one sixth of its weight of alum, dissolved in sufficient water to keep the goods wet while boiling. Then, right from the alum bath, without rinsing, plunge it into a strong preparation of black oak. That is, pretty finely bruised black oak bark . . . a pretty, bright, lively yellow will be produced."[4] Various species of oaks were used for lumber, fuel, charcoal, soap making, cabinets, furniture, posts, flooring, railroad ties, cannon carriages, and mine timbers.[5]

Medicinally, oak derivatives were valued in the treatment of diverse ailments. Writing of the lack of antiseptics during the Civil War, one physician remarked afterward, "In fact, I had but little else at my command except the cold-water dressing for wounds. From experiment I learned to improve on the plain old method, as I think, by employing a decoction of red-oak bark added to the water, which acted as a disinfectant, and by its stimulating and astringent properties promoted the healing process."[6] Powdered black oak bark was used to treat "chronic hysteria," diarrhea, rheumatism, tuberculosis, and asthma. Mixed with hog's lard it was considered a remedy for hemorrhoids and prolapsed uterus. An injected decoction of white oak bark allegedly benefited gonorrhea patients.[7]

Oak was the primary timber used in the frames of Civil War ships. The Union used white oak predominantly because it grew closer to northern shipyards. Live oak (*Q. virginiana*), considered the strongest and most durable of oaks, grew in southern states and was the most important shipbuilding timber for the Confederate navy. A critical problem for the South was a dearth of dried oak timbers. Oak takes one to three years to air dry properly to the point that shrinkage is not an issue. Lacking time, the hard-pressed Confederates were often forced to build ships with green timbers, resulting in continuously leaking vessels.[8] The frameworks of ships were not the only components made of oak. Examples include the Union timberclad gunboats *Lexington*, *Tyler*, and *Conestoga*, which had five-inch-thick oak bulwarks to protect the gun crews from small-arms fire. An iron-sheathed, four-inch oak beam was added to the bow of the CSS *General Bragg* as an improvised ram.[9] Even when the North decided to build an ironclad navy in 1862, Admiral Porter wrote that the contract for twenty single-turreted monitors called for "side-armor [that] was five inches thick fastened to a three-foot oak backing, and the turrets of eleven one-inch plates, bolted together with all the skill and ingenuity American mechanics were capable of."[10]

Civil War accounts often mention oak as shade, shelter, or firewood and less commonly as food or medicine.

Confederate nurse Kate Cumming at a hospital in Cherokee Springs, Georgia, on Sunday, Aug. 16, 1863: "To-day Dr. Quintard preached twice. As our chapel is not yet up, he had service under a large oak-tree."[11]

Major James A. Connolly, 123rd Illinois Infantry, writing to his wife on June 14, 1864, near Big Shanty, Georgia: "I received to-day, your letter mailed June 6th, and putting a large oak tree between myself and the rebel bullets, I read it with great pleasure."[12]

William L. Nugent, 28th Mississippi Cavalry, near Brandon, Mississippi, in a letter to his wife on July 28, 1863: "Like our brethren of African descent we doze away the hours of noon under the shade of the broad oaks around us on a blanket, utterly oblivious of the morrow."[13]

Sergeant Edwin H. Fay, Minden [Louisiana] Rangers, at Decatur, Alabama, in a letter to his wife on May 24, 1863: "I slept under the shade of a large oak and dreamed of you, my best beloved but only waked to find that you were not lying upon my arm and it was not your well remembered arm lying so affectionately upon my breast. No, it was only a dream."[14]

Private Wilbur Fisk, 2nd Vermont Volunteers, near Brandy Station, Virginia, in a letter to his hometown newspaper on Nov. 20, 1863: "The camp is in what was a short time ago a magnificent oak forest, but many of the tall trees had been culled out by the rebels, and now we have nearly destroyed the remainder. A continual pecking is heard all through the camp every day, and especially during this cold weather. Every axe and every hatchet that we could borrow, by fair means or foul, were put to their busiest use. The crashing of falling oaks was constant."[15]

Reuben A. Pierson, 9th Louisiana Infantry, in a letter to his father from near Camp Florida, Virginia, on Nov. 12, 1861: "Wood is quite scarce and in a few days there will not be a single oak tree in sight of camps; we will have to haul wood a mile or more."[16]

John B. Jones, clerk to the Confederate secretary of war, in Richmond, Virginia, on Aug. 8, 1863: "Oak-wood is selling to-day for $35 per cord."[17]

Captain James C. Bates, 9th Texas Cavalry, at Moorsville, Mississippi, in a letter to his future wife on June 17, 1862: "I have been laying in the shade of a great old oak in front—not of my tent, but of where it would be if I had one—dreaming day dreams and building air castles, and wondering where the star of my fortune will take me next."[18]

Captain Theodore A. Dodge, 101st New York Volunteers, near Falmouth, Virginia, on Jan. 23, 1863: "We have got out of wood today, on account of most of the rails round the farms here have been taken to corduroy the roads, and we cut down an immense oak tree close by our tents. It was a tree several hundred years old probably & very large. We got a corporal of Co. K, who is a woodman by trade, to cut it down, and we watched with great interest to see whether he would fell it clear of the Colonel's tent and ours, which he succeeded in doing. The old tree fell with a terrible crash and thousands of branches flew up into the air like spray from a waterfall."[19]

Sergeant Alexander Downing, 11th Iowa Infantry, at Grand Junction, Tennessee, on Nov. 17, 1862: "We raised the tents from the ground about three feet, by digging trenches and setting staves which we made from the red oak trees growing so plentifully here. Then we elevated our bunks about eighteen inches from the ground with the staves . . . [to prevent flooding]."[20]

Private Henry R. Berkeley, Amherst Battery in the Army of Northern Virginia, in Orange County, Virginia, on April 13, 1864: "Worked at sawmill all day, carrying off heavy oak planks; got back to camp that night very tired."[21]

Private Harvey Reid, 22nd Wisconsin Volunteer Infantry, in a letter from Danville, Kentucky, on Jan. 6, 1863: "Christmas day I was on duty chopping down white oaks, and New Years Day, I again enjoyed the pleasures of walking backward and forward on a path 4 rods long for 6 hours out of 24."[22]

Lieutenant James C. Bates, 9th Texas Cavalry, at Salt Creek, Indian Territory, on Nov. 22, 1861: "Found some large over cap acorns today which we devoured like half starved hogs. made a very good dinner today on roasted acorns & beef. Supped on pecans & beef."[23]

One recipe published during the Civil War gave instructions on how to substitute acorns for coffee: "Take sound ripe acorns, wash them in the shell, dry them, and parch until they open, take the shell off, roast with a little bacon fat, and you will have a splendid cup of coffee."[24]

Captain Theodore A. Dodge, 101st New York Volunteers, on the road to Gettysburg, Pennsylvania, on June 20, 1863: "He [a former slave] is a capital horse doctor . . . he cured my mare completely. 'That mare wants some bark,' says old Ned; so he went into the woods, cut some bark off the north side of a red oak, and dried it by the fire enough to pound it up and gave it to the mare in a feed of oats. She was quite well half an hour afterwards."[25]

Lieutenant Sidney Carter, 14th South Carolina Volunteers, in a letter to his wife from near Berryville, Virginia, on Feb. 20, 1863: "Poor fellow, he is dead—caused from vaccination. I never saw such an arm as he had, and as soon as I found out he was so bad off I turned to doctor him, cut his coat and shirt off and had Paul make a red oak poultice. Some of them sent for a Dr. and he wouldn't let the poultice go on."[26]

Oaks were the most common species of plant mentioned when describing the impacts of weapons on a battlefield.

Confederate nurse Kate Cumming in Newnan, Georgia, on May 5, 1865: "From Jonesboro to Atlanta was one scene of desolation. . . . The woods showed how desperately each side had fought for mastery. Large oaks were riven asunder, their branches broken, and scattered all around."[27]

Sergeant Edwin H. Fay, Minden [Louisiana] Rangers, at Clinton, Mississippi, in a letter to his wife on Nov. 16, 1863, as he describes the months-old Champion Hill battlefield: "Even large white oak trees are killed by Minnie balls. I saw one that I estimated had 300 in it."[28]

Lieutenant Charles B. Haydon, 2nd Michigan Infantry, near Manassas, Virginia, on Aug. 29, 1862: "I had a hearty laugh at Lt. Barden's expense. A solid 10 lb. shot struck a large black oak, abt 10 feet from the ground, under which he was lying. The splinters (the size of stove wood) knocked off his cap & battered his back smartly but without doing much damage."[29]

Assistant Surgeon William Child, 5th New Hampshire Volunteers, in a letter to his wife from Turkey Bend, Virginia, on July 29, 1864: "Those gun boat shells are terrific—awful. I have seen oak trees 15 inches t[h]rough cut off by them."[30]

Overall, oaks were mentioned most as components of the habitats frequented by the writers.

Sergeant Taylor Peirce, 22nd Iowa Infantry, writing to his wife on Nov. 1, 1863, at New Iberia, Louisiana: "We are encamped in a live oak grove near the Bayou Tash [Teche] and . . . the boys has built themselves little Brick houses and covered them with their gum blankets and have things quite comfortable."[31]

Lieutenant Charles B. Haydon, 2nd Michigan Infantry, at Hampton Roads, Virginia, on March 21, 1862: "We here see the live oak for the first time."[32]

James T. Ayers, 129th Illinois Volunteers, in Savannah, Georgia, on Feb. 5, 1865: "Those Live oaks are Shaped similar to the Apple tree only they grow quiet Larg Spreading there Lims far Around. They Don't grow high nor more than from 8 to 12 feet untill they Branch out Just like the Apple tree. Many of those trees hang full of this moss and they being evergreen makes the sight most Lovely."[33]

Private Harvey Reid, 22nd Wisconsin Volunteer Infantry, near Millen, Georgia, on Dec. 3, 1864: "The church is surrounded with trees that we suppose to be the 'live oak.' The negroes call them 'water oak.' They bear no resemblance to oak, however—have a glossy green leaf like the willow which remains green all winter."[34]

Refugee Kate Stone fleeing to Texas writes from Bellevue, Louisiana, on June 22, 1863: "We are resting for dinner in a thicket of blackjack [oak] and towering pines after a wearisome ride over the worst roads."[35]

Assistant Surgeon Dr. Daniel M. Holt, 121st New York, near Frederick City, Maryland, on August 2, 1864: "Start at 4 1/2 A.M. Orders countermanded and we encamp for the day about a mile from last night's encampment in a grove of white oaks."[36]

Lieutenant John Q. A. Campbell, 5th Iowa Infantry, near Gilbertsboro, Alabama, on Nov. 5, 1863: "The soil along the route is very rich and the timber excellent—Oak, Gum, Poplar, Hickory, Hackberry, Beech, and Walnut—Pine has disappeared."[37]

Sergeant Alexander Chisholm, 116th Pennsylvania Volunteers, in a letter to his father from near Burkesville Junction, Virginia, on April 20, 1865: "The Country up here is fine. The Apple Trees are about to bloom. There is a good deal of Oak timber in this part. We are 53 miles from Petersburgh, and about 50 from Richmond."[38]

Mary Boykin Chesnut near Camden, South Carolina, on May 8, 1864: "It is so lovely here [Mulberry Plantation] in the spring among the giants of the forest; the primeval oaks, water oaks, live oaks, willow oaks such as I have not seen since I left here. And the flowers, violets, roses, yellow jasmine; the air is laden with perfume."[39]

Lieutenant Rufus Kinsley, 74th United States Colored Infantry, in a letter to his sister from Cat Island, Mississippi, on Jan. 22, 1865: "My buildings are pleasantly situated in a live oak grove, a few rods from the water."[40]

Lieutenant James C. Bates, 9th Texas Cavalry, near Boggy Depot, Indian Territory, on Nov. 2, 1861: "Passed over some very fine country—mostly prairie and sandy post oak. We are now camped on Boggy river 1 mile from Boggy."[41]

Lieutenant Samuel T. Foster, 24th Texas Cavalry (dismounted), near Arkansas Post, Arkansas, in early October 1862: "We travel over low flat prairie country very rich but too flat to drain off the water. There is some timber—small post oak."[42]

Captain Samuel T. Foster, 24th Texas Cavalry (dismounted), near Resaca, Georgia, on May 13, 1864: "the skirmishers . . . went forward over very rough ground for about 1½ miles, where we got close enough to the Yanks in some thick Black Jack [oak] woods to hear them talking and laughing and sin[g]ing &c but they did not see us."[43]

Captain William Vermilion, 36th Iowa Infantry, in a letter to his wife from Little Rock, Arkansas, on Oct. 14, 1863: "The site [Little Rock] had once been covered thickly with an oak grove. The trees still stand around all the fine dwellings."[44]

Assistant Surgeon William Child, 5th New Hampshire Volunteers, in a letter to his wife from Newport News, Virginia, on Aug. 23, 1862: "Imagine to yourself a level of field of perhaps fifty acres. On one side a forest of oak and pine—on the other the James River. On this field are encamped perhaps twenty thousand men."[45]

Private George A. Remley, 22nd Iowa Volunteers, in a letter to his father from Rolla, Missouri, on Sept. 24, 1862: "The country through which we passed in the day time is mostly covered with forests of white & black oak, walnut, elm, and a kind of oak called the 'black jack,' extending as far as I could see on both sides of the road."[46]

Sergeant Alexander Downing, 11th Iowa Infantry, near Pittsburg Landing, Tennessee, on March 31, 1862: "The country around here is quite rough and the soil is very poor. There is a great deal of gravel and there are some rocks, but the soil works very easily. The timber here is mostly white oak."[47]

PALMETTO

"From this noble and characteristic tree is derived the well known armorial emblem on the escutcheon of the State of South Carolina."[1] The plant responsible for South Carolina's emblem and nickname, the Palmetto State, is a type of palm known scientifically as *Sabal palmetto.* The cabbage palm is the state tree of both Florida and South Carolina. With fan-shaped leaves four to six feet long that grow on tall straight trunks to eighty feet tall, this plant is usually restricted to the southeastern coast of the United States. Another species of palmetto, *Sabal minor,* is much smaller, with only the characteristic leaves above ground, the stem being buried. It grows inland from the Gulf Coast in moist areas. Civil War soldiers also came into contact with saw palmetto *(Serenoa repens)*, a plant similar in range to cabbage palm but noted for its sharp, rigid spines on the leaf stalks. During the war there was a regiment known as the Palmetto Sharpshooters, a Palmetto Battalion Light Artillery, and the Confederate ironclad, *Palmetto State.*

Confederate surgeon Francis Porcher wrote that palmetto leaves were used to make hats, baskets, mats, and thatching, and the "logs," which do not splinter, were used in the construction of forts, wharves, conduits, and other underwater structures. In his treatise he quotes a newspaper correspondent from Georgia: "You speak of black moss for mattresses. Our common saw palmetto leaves, when split into shreds with a fork or hackle, boiled, and dried in the sun one or two days, makes a light, clean, healthy, and durable mattress. Let me suggest that palmetto pillows would be cheap and comfortable for our soldiers on the coast; their corn and flour sacks would in the absence of anything better furnish ready-made pillow ticks. Our negroes are busily employed in making light, durable, and handsome palmetto hats for our soldiers—quite a protection from the sun's burning rays in the heavy drills of this and the next two months."[2]

Southern women deprived of store-bought goods operated their domestic millineries and met other household needs with palmetto leaves as raw products.

Southern refugee Fannie Cashman in Morehouse Parish, Louisiana: "The seclusion and inaccessability of the place made it difficult to obtain very elaborate wearing apparel. Palmetto grew abundantly and luxuriantly around our home, and we became expert in weaving it into hats which were very pretty and unique. The palmetto was gathered and then boiled. The boiling process bleached it perfectly white, and made it soft and pliable, thus adapted to the use we made of it."[3]

Sarah Wadley, daughter of the supervisor of Confederate railroads, near Trenton, Louisiana, on Dec. 9, 1862: "Julia invited me to go home and spend the night with her, I went . . . and learned how to plait palmetto for hats, Maggie was plaiting some for a hat for one of the negroes."[4]

Confederate nurse Kate Cumming in Mobile, Alabama, on Jan. 5, 1865: "Gentlemen's and ladies' hats are made out of saw palmetto. The ladies braid it, and use it to trim their dresses, and it makes a very pretty trimming."[5]

Kate Stone, Brokenburn Plantation near Milliken's Bend, Louisiana, on Aug. 28, 1861: "Plaiting palmetto for baskets has been the rage for several days. Jimmy and I made one for him to carry for muscadines and persimmons."[6]

Soldiers and other Civil War participants often commented on the unique, tropical appearance of palmetto plants.

Lieutenant John Q. A. Campbell, 5th Iowa Infantry, near Greenwood, Mississippi, on April 3, 1863: "The palm plant grows 'very thick' in the woods where we are camped. In the manner of its growth, the shape and size of the leaves, and length and shape of the leaf stems they are similar to the rhubarb, or 'pie plant.'"[7]

Colonel Thomas W. Higginson, 1st South Carolina [African American] Volunteers, near Beaufort, South Carolina, on Nov. 27, 1862: "Below, the sandy soil, scantly covered with grass, bristles with sharp palmettos & aloes & here & there a magnolia—all the vegetation shining & nothing soft or delicate in its texture."[8]

Sarah Wadley, daughter of the supervisor of Confederate railroads, as a refugee traveling through the Boeuf River Swamp of northeastern Louisiana, on Sept. 25, 1863: "For a mile the road was a beautiful avenue through this forest, then immediately the character of the scene changed, the large beautiful trees were still there, but around their roots the palmetto grew thick, one who has never seen it can have no conception of the effect, the scene was tropical indeed, from the forest we emerged into an open space covered thick with the glossy dark green fans of palmetto."[9]

Sergeant Lycurgus Remley, 22nd Iowa Volunteers, in a letter to his father from Milliken's Bend, Louisiana, on April 11, 1863: "Palms, from which the palm leaf fans are obtained, grow down here."[10]

Major Sherod Hunter, Baylor's (Texas) Cavalry, at Brashear City, Louisiana, on June 26, 1863: "We were again delayed here a short time in finding a road, but succeeded at length in finding a trail that led us by a circuitous route through a palmetto swamp, some 2 miles across, through which I could only move in single file."[11]

Soldiers, too, used the palmetto plant for a variety of purposes.

GENERAL ORDERS, HDQRS. DEPARTMENT OF THE SOUTH, Numbers 80. Hilton Head, S.C., June 6, 1864. By command of Major General J. G. Foster: W. L. M. BURGER, Assistant Adjutant-General. "Great care must be taken in the construction of proper sinks [latrines], which must be screened by pine or palmetto branches, and the debris covered every morning with at least 6 inches of sand. Sinks will be dug on different sides of the camps, and it will be the duty of the camp police to see that only those on the lee side of the camp are used."[12]

Sergeant William D. Dixon, [Savannah] Republican Blues, St. Catherine's Island, Georgia, on Oct. 24, 1861: "I have been hard at work putting the Palmetto roof on our shed but did not finish it. It is pretty hard to do as the Palmetto cuts the hands up badly."[13]

Lieutenant Richard Caddell, 11th Wisconsin Infantry at Brashear City, Louisiana, on Jan. 22, 1865. "I came to a small bayou which connects with Bayou Groesbeck. About sixty yards up this bayou I found a small palmetto tent which evidently had been occupied by two men. In it was a small anchor for sinking torpedoes, which I sunk in the bayou."[14]

Colonel W. W. H. Davis, 104th Pennsylvania Volunteers, at Morris Island, South Carolina, on April 6, 1864: "The southeast corner of the fort [Sumter] has been strengthened with sand bags and palmetto, and the bomb-proof at the southwest corner with sand and sand bags."[15]

Private John Westervelt, 1st New York Volunteer Engineer Corps, at Folly Island, South Carolina, on April 5, 1863: "We now took some palmetto leaves, some of which are as large as the top of a table, and put them in the hole and by pressing them against the top and sides formed an arch making a neat and comfortable place to sleep."[16]

The fruit (drupes) of palmetto was considered edible only by those accustomed to the habit, but many relished a certain vegetative part of cabbage palms.[17]

Private John Westervelt, 1st New York Volunteer Engineer Corps, at Folly Island, South Carolina, on April 15, 1863: "To day I tried something new. In the top of the palmetto grows a substance which the negroes call cabbage, and it certainly tastes much like it. I like it much better. It can be used all the different ways the same as cabbage. I use it mostly raw cut up with vinegar. One tree furnishes about as much as a small cabbage head and makes a meal for three. The palmetto is a usefull tree to us. We use the trunks to build forts. The cabbage to eat, the leaves to build huts, and the roots which grow in an enormous mop, to tie them on with."[18]

PERSIMMON

Common persimmon (*Diospyros virginiana*) is a member of the ebony family, known for its hard wood. This small to medium-sized tree is found throughout the eastern United States in a variety of habitats and grows largest in rich, moist soils. Once sought for its fruit, persimmon bears a type of orange to reddish-purple berry about one and one-half inches in diameter. The flesh of the fruit is very astringent if picked before fully ripened, leading to the adage that it should not be eaten until after the first hard frost of autumn. Many types of wildlife also eat persimmon fruits, and the seeds seem to germinate best after passing through the digestive tract of a mammal. Opossums and raccoons are especially adept at "sowing" persimmon seeds.

The dense, hard wood of persimmon was used indirectly in the outfitting of soldiers and citizens during the Civil War. Shoe lasts on which boots and shoes were manufactured were made of persimmon, as were the shuttles of looms that weaved the cloth of uniforms. Buttons were made from persimmon seeds. Some tool handles and fifes were also made from persimmon wood.[1]

Confederate surgeon Francis Porcher stated that, as a medicinal plant, "The persimmon should be used in camps as an astringent" (a material that draws tissues together), and he claimed that the fruit, fresh or dried, was very valuable in the treatment of diarrhea, chronic dysentery, and uterine hemorrhage.[2] After the war General "Stonewall" Jackson's former medical director, Dr. H. H. McGuire, referring to the lack of medical supplies available to Confederate surgeons wrote, "His scanty supply of medicines and hospital stores made him fertile in expedients of every kind. I have seen him search field and forest for plants and flowers, whose medicinal virtues he understood and could use. The pliant bark of a tree made for him a good tourniquet; the juice of the green persimmon, a styptic [used to stop blood flow]; a knitting needle, with its point sharply bent, a tenaculum, and a pen-knife in his hand, a scalpel and bistoury."[3]

"Persimmon Regiment" was a derogatory term applied to units accused of breaking ranks to gather the fruits. One regiment, the 35th Ohio Infantry, acquired the brand when fifteen of its men were captured after straying to search for persimmons in December 1861.[4] Such criticism obviously deterred very few hungry soldiers.

Lieutenant Charles B. Haydon, 2nd Michigan Infantry, near Washington, on Sept. 23, 1861: "There was a persimmon tree loaded with rich fruit abt 4 rods in front of the line. Its fruit had often been coveted by our men. I concluded to go & get some. I was busily knocking them off with a pole when a rascal fired at me, the ball striking abt 20 feet short & a little to one side. I grabbed up my hands full of persimmons & made no unnecessary delay in returning inside the lines to my proper place."[5]

Major James A. Connolly, 123rd Illinois Infantry, near the Ogeechee River, Georgia, Nov. 27, 1864: "Persimmons grow by the roadside in abundance; our orderlies gather them in their handkerchiefs as they ride along, and bring up to us, so that we just ride along and eat persimmons, until we are almost tired of them. They are much finer than I ever supposed they were; tasting very much like excellent figs; I have eaten persimmons in Illinois, but they are very little like the persimmons of this part of Georgia."[6]

John B. Jones, clerk to the Confederate secretary of war, in Richmond, Virginia, on Nov. 19, 1864: "We had a delicious treat of persimmons to-night—a quart bought for a dollar. They were delicious, and we enjoyed them hugely."[7]

John S. Jackman, 9th Kentucky Infantry, near Chattanooga, Tennessee, on Nov. 1, 1863: "Sunday. Rode out 5 or 6 miles hunting black-haws and persimmons. Nice day."[8]

Captain Jacob Ritner, 25th Iowa Infantry, in a letter to his wife from Stevenson, Alabama, on Nov. 17, 1863: "For the last three or four days we have been among the Cumberland Mountains. I never saw such nice clear springs and running streams. We had an abundance of the best water and plenty of beechnuts and persimmons, which is about all this country produces."[9]

Brigadier General Alfred Pleasonton, Headquarters Cavalry Brigade, in a report from northern Virginia on Dec. 5, 1862: "The two deserters from Fifteenth Virginia Cavalry report there have been no rations issued them for three weeks, and for the last three days they have lived on berries and persimmons; that their horses are in wretched condition, and that the rebel soldiers are discontented for want of pay, six months' being due them. These men were volunteers."[10]

Private Allen M. Geer, 20th Illinois Volunteers, at Jackson, Tennessee, on Oct. 7, 1862: "Took another gay swimming excursion with some jovial comrades. Found an abundance of persimmons in our tramp."[11]

Some wrote of persimmons in a lighter vein.

Robert D. Patrick, 4th Louisiana Infantry, at Enterprise, Mississippi, on Aug. 12, 1863: "When I went in, I found a gaunt, sallow faced Tennesseean sitting in the little parlor. He looked like he had bean reared from infancy on goobers, green apples, persimmons and dirt."[12]

Sergeant Henry C. Lyon, 34th New York Volunteer Infantry, near Poolesville, Maryland, in a letter to his brother on Feb. 2, 1862: [in reply to his brother's questions about seeds that were mailed home] "The largest of the animals are called Paw-Paws the others are just 'Old Persimmons' those are the names. Plant them in the Spring. I will be up there by the time they are large enough to make a shade for me to lie under."[13]

Private Nelson Stauffer, 63rd Illinois Infantry, near Beaufort, South Carolina, on Feb. 1, 1865: "Drew Codfish instead of bacon, we couldn't cook them so that they could be eaten but we had a good deal of fun with them. We nailed them to a persimon [*sic*] tree from the trunk almost to the top. Then made a great many remarks about the peculier odor of the Southern Aristocracy growing on persimmon trees."[14]

Beer can be made from persimmons and, like alcohol in any form, was a cherished commodity. One Civil War recipe for the brew follows:

"To make Persimmon Beer—Gather the persimmons perfectly ripe and free from any roughness. Work them into large loaves with bran enough to make them consistent; bake them so thoroughly that the cake may be brown and dry throughout, but not burned. They are then fit for use. But if you keep them any time it will be necessary to dry them frequently in an oven moderately warm. Of these loaves broken into a coarse powder, take eight bushels. Pour on them forty gallons of cold water, and after two or three days draw it off; boil it as other beer, adding a little hops. This makes a very strong beer."[15]

Major James A. Connolly, 123rd Illinois Infantry, near Allatoona, Georgia, Oct. 10, 1864: "I saw a light in a house by the road side, and being very thirsty, stopped to get a drink of water; they told me they had no water, but could give me a drink of 'persimmon beer,' something I had often heard of, but had never seen or tasted. I gladly accepted it, but it was poor stuff—tasted about like vinegar diluted with water, and colorless like water."[16]

PINE

Very soon after his assassination President Lincoln was placed in a pine (*Pinus* spp.) coffin.[1] From the cradle, where the umbilical cords of infants were dabbed with turpentine, to the grave no group of wild plants were more utilized by humans during the Civil War than pines. Thirteen species of pines grow in the eastern United States in a variety of habitats. In those states that remained in the Union, eastern white pine (*P. strobus*) was most important, while the coastal region of the Carolinas and Deep South was blanketed with vast coniferous forests of longleaf (*P. palustris*), shortleaf (*P. echinata*), loblolly (*P. taeda*), and slash (*P. elliotti*) pines.

From the beginning of the first New England colonies until the twentieth century, eastern white pine was the main source of softwood timber in the country. As a building product with a host of uses, lumber from white pine was used in cabinetry, matches, siding, framing, doors, furniture, and boxes.[2] "Passing boxes" were boxes made from white pine that were used in passing out cartridges or moving ammunition from magazines to guns.[3] The long straight trunks of white pine, one of the tallest trees in eastern North America, were especially prized for ship masts. Although much less resinous than the southern pines, eastern white pine yielded some products such as turpentine and tar.

Longleaf pine, named for its needles up to eighteen inches long and also called pitch pine, is the most resinous of eastern pines, a trait that made it America's leading source of naval stores for two hundred years. Tar and pitch were vital to the construction and maintenance of merchant fleets and navies of the era. Tar was extracted by slowly burning resinous pine wood, usually in a kiln, and pitch was made by boiling down tar in a large kettle. Tar was used for waterproofing ropes in critical running and standing riggings, and in hawsers to moor the ships and support the anchors. Hulls were made watertight with a coating of pitch that also provided protection against wood-boring marine mollusks, a scourge of mariners since the first wooden ships. At the beginning of the Civil War, North Carolina was the largest producer of naval stores in the world. The term *tar heel* is alleged to have originated when, during a battle, a group of North Carolina soldiers failed to hold a hill and retreated in disarray. A Mississippi unit chided them for forgetting to tar their heels that morning.

The longleaf pine forests of North Carolina also produced the most turpentine and rosin. Turpentining involved chopping notches into living trees and collecting resin that flowed from the wounds in boxes below. When the resin was distilled it yielded liquid turpentine and rosin. The uses of turpentine were almost limitless in number and creativity. Medicinally, it was used as a stimulant, diuretic, antiseptic, and laxative. At various times turpentine was used in the treatment of worm infestations, epilepsy, tetanus, diabetes, yellow fever, typhus, tuberculosis, rheumatism, toothache, dysentery, and puncture wounds. Surgeons during the heat of naval battles injected hot turpentine directly into wounds and coated stumps resulting from amputations with it. The treatment of many horse and mule ailments involved turpentine. Aside from medicine, turpentine was used in making soap, candles, lighting oil, furniture polish, and rodent repellants. Confederate soldiers waterproofed their boots with a concoction of beef and hog tallow, beeswax, and turpentine. Rosin was also used in waterproofing and soap making. A garnet-colored dye was made from pine roots. The wood of longleaf pine and to a lesser extent other southern pine species was important for all types of general construction, railroad cars and crossties, charcoal, fuel, and shipbuilding. Southern shipbuilders used longleaf for masts, spars, planking, and decking. Pine buoys were used to float rope obstructions in Charleston harbor. Both the Union and Confederate governments considered pine products important sources of revenue as items of export. The North routinely exported pine naval stores during the war, and the blockaded South made frequently unsuccessful efforts. The Confederate schooner *Adelaide,* bound for Nova Scotia, had a cargo of six hundred barrels of turpentine when the USS *Ellis* ran her aground off the North Carolina coast in October 1862. The British sloop *Anna Eliza* wrecked soon after departing on the South Carolina coast with ten thousand gallons of turpentine spirits in May 1864.[4]

Among the many uses of pines, Civil War soldiers readily exploited the various species at hand for shelter, furniture, fuel, lighting, and even landscaping.

Private William G. Bentley, 104th Ohio Volunteer Infantry, in a letter to his family from Knoxville, Tennessee, on Oct. 23, 1863: "We are building barracks for winter quarters. They are built of pine logs and will be very comfortable when we get them finished. They are 10 ft. wide by 14 ft. long & about 6 ft. high with a fire place in each."[5]

Private Wilbur Fisk, 2nd Vermont Volunteers, in camp near White Oak Church, Virginia, in a letter to his hometown newspaper on April 13, 1863: "Regular streets were laid out for each company in the different regiments of this brigade. . . . We brought pine trees from the woods and planted a row on each side of every one of these, making each street a vista, at once pleasing and beautiful. The officers had their tents encircled by a grove of pine, and many of the men did the same."[6]

Private Henry R. Berkeley, Amherst Battery in the Army of Northern Virginia, near Fredericksburg, Virginia, on Nov. 24, 1862: "We fixed up our bed off the ground on some pine poles, putting little pine twigs a foot deep on top of the poles and some hen's nest grass on the top of the twigs. This made a very good and comfortable bed."[7]

Colonel Francis C. Barlow, 61st New York Volunteers, in a letter to his brother from Yorktown, Virginia, on April 18, 1862: "Tell R that I got my tent covered with pine limbs this afternoon to make it cool & have had a thick floor of pine branches & it is quite clean."[8]

Captain William J. Seymour, 1st Louisiana Brigade, near Rappahannock Station, Virginia, on Dec. 3, 1863: "In our Brigade there are two hundred and fifty men who have neither blankets nor overcoats. It is a great wonder that these men do not freeze to death these terribly cold nights. Many of them use pine leaves and boughs wherewith to shield them from the cold, while others sit up by the fire all night."[9]

Lieutenant Edmund D. Patterson, 9th Alabama Infantry, near Fredericksburg, Virginia, on Feb. 23, 1863: "We have made ourselves very comfortable here; most of the boys have built little huts of pine poles and roofed them over with pieces of tent cloth."[10]

Assistant Surgeon Dr. Daniel M. Holt, 121st New York, near Petersburg, Virginia, on July 8, 1864: "Hot as it can be. No one to blame that I know of, but give the Quarter Master and all hands a blowing up because I have no shade trees drawn to break off the sun from my quarters. I get what I blow about. Three army wagons and a gang of hands are at work setting out pine trees all around my quarters! It does do good sometimes to express a righteous indignation at unbearable evils."[11]

Private Richard H. Brooks, 51st Georgia Infantry, in a letter to his wife from Fredericksburg, Virginia, on April 27, 1863: "an this is one Long cold winter I have spent without eaven seeing a Light wood not. We get nothing to burn but green pine, oak, cedar, an chestnut."[12]

Private John M. King of the 92nd Illinois writes on July 22, 1864, near Resaca, Georgia: "I stayed at the house of a citizen that night. . . . In the evening the husband brought in an armful of pitch pine cut and split into bits about eighteen inches long and the size of a broom handle. . . . the wife drew the two 'handirons' of the fireplace together . . . where she placed one end of the stick of pitch pine, the other end resting on the hearth . . . the pitch pine took fire at once and sent a glaring light through the house. . . . One could see to read in most any part of the room. . . . the wife asked me if we had plenty of pine in Illinois. When I told her we had none . . . she asked, 'What on earth do you'uns do for lights in the evenin's?' I told her we used candles . . . 'Candles! [she said] What do they look like?'"[13]

Civilians, especially those in the economically depressed South, also exploited pine products opportunistically.

Confederate refugee Sarah Morgan Dawson on Linwood Plantation near Port Hudson, Louisiana, on Nov. 4, 1862: "Inside [the sugar house], it was lighted up with Confederate gas, in other words, pine torches, which shed a delightful light, neither too much nor too little."[14]

Confederate nurse Kate Cumming in Mobile, Alabama, on Feb. 3, 1863: "I have been not a little amused at the novel lights we have; instead of oil and candles, nature has bountifully supplied us with illuminators in our pitch pine knots. . . . We put some pieces of pine in the grate, which gives light enough to see each other; all we can do is converse, as it is impossible to sew or read by this light. We are compelled to retire in the dark, or else run the risk of having our complexion and every thing else ruined by the smoke of the pine torches. These are things which every body laughs at, saying it is war times, and they will soon be over."[15]

John B. Jones, clerk to the Confederate secretary of war, in Richmond, Virginia, on Feb. 4, 1865: "The City Council is having green 'old field pine' wood brought in on the Fredericksburg railroad, to sell to citizens at $80 per cord—a speculation."[16]

Confederate sympathizer Myra Inman at Cleveland, Tennessee, on Oct. 20, 1863: "Mother and sister have gone out to the woods to get some pine resin for us to take to cure the chills."[17]

Confederate refugee Belle Edmondson near Serepta, Mississippi, on Oct. 17, 1864: "Well! here I sit to Night 20 miles from Pontotoc . . . fortunately met with comfortable quarters for the Night . . . I have a clean bed, and nice pine torch."[18]

Sarah Wadley, daughter of the supervisor of Confederate railroads, near Trenton, Louisiana, on Dec. 31, 1861: "We had a very pleasant Christmas; the day after Christmas day, Miss Mary and I fixed up a little pine tree as a Christmas tree, we had no costly gifts, but a few sugar plums in lace bags, and some home made Cornucopias with two or three little wax candles made the tree very attractive to the children."[19]

Pine's role in the implements of war and its involvement in danger and death was ubiquitous.

Private Isaac Jackson, 83rd Ohio Volunteer Infantry, near Mobile, Alabama, on April 13, 1865, writing of the recent fall of Spanish Fort: "I have given you a rough sketch of the works. The 2 & 3 abattis were of short stubby pine limbs. The 3rd one had long sharpened stakes in it, with a wire stretched in front of it for the purpose of throwing the men on the stakes."[20]

Frederic A. James, U.S. Navy, describing his surroundings on his first day as a prisoner of war at Andersonville Prison, Georgia, on June 1, 1864: "The lot contains 18 acres & is fenced in with a solid stockade of hard pine logs about 10 or 12 feet high. . . . Between 19 & 20,000 prisoners are now here."[21]

Private Gottfried Rentschler, 6th Kentucky Volunteer Infantry [Union], in a letter to the *Louisville Anzeiger* from near Atlanta, Georgia, on July 25, 1864: "Our brigade headquarters was located on a mountain ridge and, on this ridge, a tremendously high pine tree stood in front of General Hazen's tent. Hazen had a ladder made, and from the top of this pine tree one could see the country in a wide circle and could see Atlanta."[22]

Sergeant Rice Bull, 123rd New York Volunteer Infantry, near Dumfries, Virginia, on Jan. 19, 1863: "To proceed, the Pioneers aided by our troops corduroyed the road. This was done by cutting the pines growing along the road and laying them across close together. This made a very rough uneven footing, but the wagons and artillery could move slowly over it."[23]

Major William J. Bolton, 51st Pennsylvania Volunteers, near Loudon, Tennessee, on Oct. 28, 1863: "They [Confederate pontoon boats] were very heavy, built of 2 1/2 inch southern pine plank, the main body of each boat being rectangular in form, and finished at the upstream, and by an addition which was equilateral triangle in plan."[24]

Private Henry R. Berkeley, Amherst Battery in the Army of Northern Virginia, on Feb. 18, 1864: "It took the train all day to run from Richmond to Gordonsville. It was crowded with soldiers and the engine had nothing but green pine wood to make steam."[25]

Private Thomas Jones, 48th New York Volunteers, in a letter to his sister from Daufuskie Island, Georgia, on March 11, 1862: "they set us to work carrying pine poles on our shoulders about the distance from one and one half to two miles. There were 20,000 poles and we carried them in nine days . . . when this was done, we had to haul eight large cannon across a marsh over a mile long [using the poles as corduroy road]."[26]

Corporal Alexander Downing, 11th Iowa Infantry, on the Ogeechee River near Savannah, Georgia, on Dec. 17, 1864: "Large details of men from our division were sent out to cut and prepare timber for the engineers to build a wharf at the landing so that the boats can be unloaded more readily. Several hundred of us were at work, some cutting the trees—tall pines, others cutting them into proper lengths, and still others hewing and squaring the timbers."[27]

Charles T. Quintard, chaplain, 1st Tennessee Infantry, at Columbus, Georgia, on April 22, 1865: "The fight for the defense of Columbus was quite a brisk affair. . . . At this time I had made no preparation for the coming of the enemy. I had at my house the money [$33,000] collected at the offertory in the morning. This Mr. Noble put in the top of a tall pine tree in the stable yard."[28]

Major James A. Connolly, 123rd Illinois Infantry, near Neuse River, North Carolina, March 21, 1865, in a letter to his wife: "While we were driving in their skirmishers yesterday I received a gentle reminder that 'a body ought to be careful' for a rebel bullet struck the limb of a pine tree some five feet above my head, and glancing downward struck the plate of my sword belt and plunged into the leaves at my feet. It didn't hurt any and I was very glad of it."[29]

Walt Whitman after the battle of Fredericksburg, December 1862: "the wounded [are] lying on the ground, lucky if their blankets are spread on layers of pine or hemlock twigs, or small leaves."[30]

Private John Westervelt, 1st New York Volunteer Engineer Corps, at Folly Island, South Carolina, on April 26, 1863: "An accident occurred to a Lieut of the ... Illinois. He was walking under a pine tree that was hollow at the bottom in which someone had built a fire. It was nearly burned off and falling and suddenly nearly crushed him to death."[31]

Lieutenant Rufus Kinsley, 74th United States Colored Infantry, in a letter to his sister from Cat Island, Mississippi, on Jan. 22, 1865: "Two or three nights ago the lightning struck and shivered into fragments a large pine tree that stood sixteen feet from my door, and at the same time, a splendid oak that stood only seven feet from the door. So you see the elements are not less dangerous than 'the enemy.'"[32]

Sergeant James H. Avery, 5th Michigan Cavalry, in eastern Virginia, on May 13, 1864: "We passed through Fair Oaks, White Oak Swamp, Chickahominy, Gaines Mill, and Seven Pines. As we marched along, we could see the tall pines, with their boughs literally trimmed, and many a one with its top cut entirely off with shot of shell. The trees were marked with balls from the ground up, as far as we could see. One can imagine that in such a storm, many men must die."[33]

Sergeant Alexander Downing, 11th Iowa Infantry, near Columbia, South Carolina, on Feb. 15, 1865: "At times when marching on a road alongside the burning pine timber, we became so blackened from the smoke as to look like negroes, while the heat from the burning pitch was frightful."[34]

Lieutenant Colonel Charles F. Johnson, 81st Pennsylvania Volunteer Infantry, in a letter to his wife from eastern Virginia on Nov. 8, 1861:

"One of the men at Nottingham shot a member of his company through the head, killing him instantly, we brought his body to camp where it now lies in a pine board box, we will take it to Washington."[35]

Private Henry R. Berkeley, Amherst Battery in the Army of Northern Virginia, at Fort Delaware Prison, on March 28, 1865: "From this little window . . . The hospital dead-house, and a shed where the plain, pine coffins are made for our dead can also be seen. It is a gloomy and sad sight, those high piles of coffins, and they carry from eight to a dozen of them every morning, filled with our poor dead boys, whose loved ones down in Dixie are anxiously watching for, and bury them on the New Jersey shore."[36]

A letter from the uncle of Private Galutia York, 114th New York Volunteer Infantry, to his family on May 16, 1863, from Berwick City, Louisiana: "a sad duty devolves on me to announce to you that Galutia is no more he died on the 14 inst at 2 oc P.M. . . . the place he is buried is a beautiful elevation on the bank of the river . . . a smooth plain pine board stands at his head bearing this inscription G. York Co. G 114 N.Y.S.V. aged 19 years."[37]

The various species of pines in the Civil War arena occupied different ecological niches. Some in old-growth form were components of a climax forest; others thrived in early or intermediate stages of plant succession. Most were dependent or flourished on fire at some point in their life cycle. Pines are among the most important plants for wildlife in the eastern United States. A host of bird species, rabbits, squirrels, mice, and other rodents eat pine seeds. Deer and grouse eat pine needles, and porcupines, beavers, and small rodents eat pine bark. The foliage provides cover for many animals, and pine needles are used as nest material by several species of songbirds. Most soldiers mentioned pines in their writings in an ecological context as a component of the habitats to which they were exposed.

Major James A. Connolly, 123rd Illinois Infantry, near Louisville, Georgia, Nov. 29, 1864: "Our headquarters are in the edge of a beautiful grove of pines. This grove was formerly cultivated land, and the marks of the furrows are as distinct now as if the land had been plowed two years ago. Yet there are many pines growing on it now that are a foot in diameter, and in most places the pines grow so thickly as to render it almost impossible to ride through them. This is the kind of land that is called 'Old Fields' in the South."[38]

Sergeant George E. Stephens, 54th Massachusetts [African American] Regiment, near Jacksonville, Florida, on March 6, 1864: "You can see nothing but pine woods, marsh, and every five or ten miles a cluster of dilapidated, deserted huts, with no sign of agricultural thriftiness. But immense tracts of this pine-woods land are prepared for the collection of pitch. The trees are tapped, and near the roots cavities are hewn out, into which the pitch collects."[39]

John S. Jackman, 9th Kentucky Infantry, near Greensburg, Louisiana, on Aug. 11, 1862: "Been another hot day. Our road led through an interminable pine forest. The country is perfectly level, and there is no undergrowth to obstruct the view—can see a long way through the woods, the pines standing not very thick and straight and beautiful. . . . What stillness in these pineries! You never see a bird, nor even hear a grass-hopper chirp."[40]

Sergeant Taylor Peirce, 22nd Iowa Infantry, writing to his wife on May 21, 1865, near Augusta, Georgia: "The land is poor. . . . The soil is sandy and generally covered with the yellow Pine where it is not cleared off and is generally level."[41]

Private William G. Bentley, 104th Ohio Volunteer Infantry, in a letter to his family from near Kingston, North Carolina, on March 12, 1865: "You have read of the pine woods of North Carolina I expect, but no one can imagine the extent of them unless they have traveled through them. We have marched nearly 100 miles in a direct line from Wilmington since Monday and have seen only 5 or 6 cleared plantations that deserve the name of cleared. The country is almost as level as a floor and grown up with pitch pine and cypress. In some places so thick that a man can scarcely walk threw them."[42]

Private John F. Brobst, 25th Wisconsin Infantry Regiment, near New Bern, North Carolina, on Feb. 21, 1865, in a letter to his future wife: "Negroes, white sand, and scrub pine constitutes what I have seen of North Carolina."[43]

Private Harvey Reid, 22nd Wisconsin Volunteer Infantry, near Milledgeville, Georgia, on Nov. 24, 1864: "We are now in the great 'pitch pine' region that extends along the whole Atlantic coast in the Southern States."[44]

John B. Jones, clerk to the Confederate secretary of war, near Wilmington, North Carolina, on May 13, 1861: "The pine woods in some places have a desolate appearance; and whole forests are dead. I thought it was caused by the scarifications for turpentine; but was told by an intelligent traveler that the devastation was produced by an insect or worm that cut the inner bark."[45]

Private Wilbur Fisk, 2nd Vermont Volunteers, near Chancellorsville, Virginia, in a letter to his hometown newspaper, on Nov. 29, 1863: "The land where we are now, is covered with small second-growth pine, and looks as if it had been under cultivation once, but probably worn out and abandoned for more fertile regions. Part of the woods that we have been through is grown up to oak and other solid timber of all sizes, and has probably been forest from time immemorial."[46]

Sergeant Rice Bull, 123rd New York Volunteer Infantry, near Middletown Station, North Carolina, on March 8, 1865: "The road was through a great pine forest of mammoth trees; it seemed as though we were passing through a tunnel. The trees were so close to the road they only left a pathway. We had reached the tar and resin producing country. The large trees were all tapped for their gum which was then taken to factories and converted into tar and resin. We passed one of these black looking factories but it was in flames."[47]

Private Nelson Stauffer, 63rd Illinois Infantry, near the Oconee River, Georgia, on Nov. 28, 1864: "Marched about 20 miles in the Pine Forest or we might call it a Wilderness. nothing but Pine trees and the pine leaves were so thick on the ground that it made very slowish marching."[48]

Lieutenant Charles B. Haydon, 2nd Michigan Infantry, near Washington, on Dec. 25, 1861: "The pines will someday, not far off, conquer

Virginia if we do not. There are the densest growth of scrub pines I ever saw on what not many years ago were cultivated fields. The pines are all scrubby & seldom go higher than 40 feet. In thickets where the trees are 6 inches through you can still see the marks of the plow & the ridges thrown up along the rows of corn can be distinctly traced. . . . There are some fields but just abandoned & in which the trees are not more than 4 feet high but just as thick as they can stand."[49]

Private Van R. Willard, 3rd Wisconsin Volunteers, near Fredericksburg, Virginia, on April 30, 1863: "This country in the rear of Fredericksburg right well deserves the name of 'The Wilderness,' for it is a wilderness in every sense of the word—one of deep, interminable forests of pine—not the tall, noble pines of the North, but low, stunted scrubs, very thick, with limbs reaching to the ground. These pines are of a second growth and so thick together that it is almost impossible to force one's way through among them. This pine wilderness is interspersed with little groves of scrub oak not less dense than the pine. This wilderness is almost unbroken."[50]

Captain Jacob Ritner, 25th Iowa Infantry, in a letter to his wife from Iuka, Mississippi, on Oct. 12, 1863: "It is a very rough, poor country, what they call 'pine barrens.' Saw some very nice tall pine timber and thousands of small pines like folks plant in their yards up North. Here they grow wild all through the woods."[51]

Private William R. Stilwell, 53rd Georgia Volunteers, in a letter to his wife from near Fredericksburg, Virginia, on March 15, 1863: "Our present camp is situated in another pine grove. . . . The pine grove extends about a mile east of our quarters and furnishes a beautiful place for one to stroll off by themselves to secret devotion and times while the sun was setting in the far west and the March winds come stealing through the pines making a solemn sound and all nature seems to be hushed in gems of pleasure[.] Here many times have I bowed to offer up my evening prayer for dear friends at home and to commit them to God's care and mercy."[52]

General Robert E. Lee, CSA, in a letter to his wife from near Manassas, Virginia, on Oct. 25, 1863: "I moved yesterday into a nice pine thicket, and Perry is to-day engaged in constructing a chimney in front of my tent, which will make it warm and comfortable."[53]

SASSAFRAS

Without sassafras (*Sassafras albidum*) General Benjamin Butler would not have been able to experience the delights of a genuine New Orleans gumbo dinner during his occupation of that city. The dried, powdered leaves known as *file* are the finishing touch to the traditional, spicy soup of the swamps. Sassafras is usually a shrub or small tree but can grow to eighty feet tall and three feet in diameter in optimum conditions. It often forms dense, shrubby thickets. The deciduous leaves are unusual in that three different shapes may grow on the same plant. Sassafras is widely distributed throughout the eastern and southern United States.

As a medicinal plant, sassafras is reported to be one of the first exported to Europe from the American colonies. Confederate surgeon Francis Porcher used the roots to brew a "warm, aromatic, mucilaginous tea" to treat fever, pneumonia, bronchitis, catarrhs, and mumps. It was also used to treat measles in the Civil War and a host of other ailments before this time.[1]

The wood of sassafras is very durable yet somewhat brittle. It was used for ox yokes, cooperage, light boats, poles, posts, and crossties. Bedsteads and roost poles in chicken houses were once made of sassafras to deter insect pests. Although an additive is manmade today, the odor of root beer drinks once derived from sassafras roots. A yellow to orange dye was also made from the roots.[2]

References to sassafras during the Civil War usually related to its use as a beverage. Safrole, an oil found in the plant, has been reported to be carcinogenic in lab animals.

Private Isaac Jackson, 83rd Ohio Volunteer Infantry, at Rocky Springs, Mississippi, on May 8, 1863: "We live fat. Plenty of the best mutton and beef. . . . But we don't have any coffee for a while, but in its stead good sassifras tea."[3]

Refugee Virginia Rockwood, referring to Uniontown, Alabama, soon after surviving the siege of Vicksburg: "I had a pleasant time there, met many of my home friends who had run away from Vicksburg on account of the terrible fighting. . . . I gave a big party. . . . Had charades and beautiful music. My party was fine. The eatables were hot rolls, scrambled eggs, boiled bacon, chicken salad, hot sassafras tea, had nice cream for the sassafras tea."[4]

Private George A. Remley, 22nd Iowa Volunteers, in a letter to his father from Rolla, Missouri, on Sept. 24, 1862: "There is 'right smart' sassafras all over the hills around here. We had tea of it once and I thought it was first rate."[5]

Kate Stone, Brokenburn Plantation near Milliken's Bend, Louisiana, on March 17, 1863: "The plums and sassafras are in full bloom and the whole yard is fragrant. We all drank sassafras tea for awhile but soon got tired of it, pretty and pink as it is."[6]

Confederate sympathizer Myra Inman at Cleveland, Tennessee, on Feb. 21, 1865: "A pretty day. We went after some sassafras roots this eve."[7]

Captain Jacob Ritner, 25th Iowa Infantry, in a letter to his wife from near the Black River in Mississippi, on Aug. 2, 1863: "We have a very poor place to camp—on a sharp ridge and 3/4 of a mile to water. However our tent happened to come in a pretty good place, under a sassafras tree that makes a fine shade."[8]

Sassafras, like a host of other plants wild and cultivated, was used to brew an alcoholic drink during the war.

"A cheap and wholesome beer for the use of soldiers, or as a table beer, is prepared from the sassafras, the ingredients being easily obtained. Take eight bottles of [sassafras] water, one quart of molasses, one pint of yeast, one tablespoonful of ginger, one and a half tablespoonful of cream of tartar, these ingredients being well stirred and mixed in an open vessel; after standing twenty-four hours the beer may be bottled, and used immediately."[9]

SPANISH MOSS

The name Spanish moss (*Tillandsia usneoides*) is a misnomer. It is not a true moss like sphagnum but rather a flowering plant in the bromeliad family closely akin to pineapples. It grows in coastal and swampy regions of the southeast and as far south as South America. Often associated with our images of southern swamps, Spanish moss grows on trees in long, draping, threadlike, gray veils where as an epiphyte it absorbs moisture and nutrients from the air. The plants are not parasitic and don't normally harm their host trees.

Many types of wildlife use Spanish moss in their life cycles. Squirrels and birds use it for nest materials. Northern parula warblers (*Parula americana*) build their nests almost exclusively in draping clumps of the plant. Some species of bats roost in Spanish moss and it is the sole habitat for one kind of jumping spider.

Humans have used Spanish moss for centuries. Early European colonists recorded Native Americans wearing clothing made from the plant. Louisiana Acadians made a concoction of mud and Spanish moss known as *bousillage* for mortar and house insulation. Later an entire commercial industry developed around the harvest and processing of the plant into manufactured products. It was used for packing materials, mulch, and in saddle blankets. Thousands of tons were ginned and used to stuff mattresses until as late as 1975, when synthetic fibers replaced the natural filaments.

Because Spanish moss receives all of its nutrients from the air, it is very sensitive to wind-borne pollutants such as heavy metals from exhaust fumes and pesticides. Early explorers often remarked about the dismal, dreary atmosphere associated with moss-laden swamps. It is now known that the presence of healthy Spanish moss is an indicator of good air quality.

Most references to Spanish moss in Civil War diaries and journals are attributable to Union soldiers due to their unfamiliarity with the plant and its novelty. Soldiers from both sides were quick to utilize the plant, especially for bedding.

Major James A. Connolly, 123rd Illinois Infantry, near Louisville, Georgia, Nov. 28, 1864: "we are now in the country where the 'Spanish Moss' begins to show itself, and General Baird tells us that we will find it still more abundant as we approach the coast. It is a parasite like the mistletoe, has a dark grayish appearance, and hangs in ringlets from the limbs, draping the trees completely, and giving them a gloomy, funereal appearance; the General says that this moss is gathered, scalded with hot water, then dried and whipped, when all this outside coating of gray flies away in dust, leaving the black, glossy curly moss used by upholsterers."[1]

Captain Charles B. Haydon, 2nd Michigan Infantry, near Vicksburg, Mississippi, on June 20, 1863: "The trees are loaded with the long grey

Southern moss which hangs from the limbs in clusters & sheets from 2 to 10 feet in length (perpendicular) and swings loose in the wind. This gives to everything a sort of dull somber appearance. It looks old, very old, as though everything was on the decline."[2]

James T. Ayers, 129th Illinois Volunteers, in Savannah, Georgia, on Feb. 5, 1865: ". . . there is A Kind of moss grows here on the Branches of the trees many of the Trees being hung full of this moss in some places all the timbers are Loaded or all there Branches. This moss Resembles in Coller that of Water Rotted Hemp toe. . . . This makes the forests have A novel, tho A beautiful appearance. Well I am toald there is great use made of this moss. It is gathered by the People and used for matrasses under beds, stuffing sofas &c., being very useful."[3]

Private Harvey Reid, 22nd Wisconsin Volunteer Infantry, near Davisborough, Georgia, on Dec. 1, 1864: "Plucked a specimen of Southern moss and also a few holly leaves. The moss is softer and heavier that I expected—more nearly like other plants. It seems to grow entirely on a single variety of oak."[4]

Sergeant Rice Bull, 123rd New York Volunteer Infantry, near Savannah, Georgia, on Dec. 10, 1864: "The land was low and damp so we gathered boughs from the pine trees and Spanish moss from the live oaks to make a foundation for our beds. There were many live oaks, beautiful to look at with their long veils of moss that trailed to the ground, seeming for the world like the beards of the Old Testament prophets as seen in Bible pictures. The moss made a good foundation for our beds."[5]

Colonel Thomas W. Higginson, 1st South Carolina [African American] Volunteers, at Beaufort, South Carolina, on Aug. 26, 1863: "I hv. had a floor to my tent & a pretty penthouse over it of poles & gray moss, which we use a good deal—this keeps off the sun, & I have scarcely ever had such comfortable quarters."[6]

Sergeant Lycurgus Remley, 22nd Iowa Volunteers, in a letter to his father from Milliken's Bend, Louisiana, on April 11, 1863: "The ground was carpeted with green grass, and the large trees (some of them being about six feet in diameter) were draped with a mossy, gray, pendent parasite, called, I believe, Spanish Moss. When dried, it looks almost like horse hair, and is used as stuffing for cushions, etc. It makes a *splendid bed,* and it is this upon which we now sleep."[7]

Private John F. Brobst, 25th Wisconsin Infantry Regiment, near New Bern, North Carolina, on Feb. 21, 1865, in a letter to his future wife: "Have been very busy building [winter] quarters and are getting along fine. My squad has got ours completed and are living in it. It is a small log hut with a cloth roof, chinked up and corked up with moss, makes it quite comfortable this time of the year."[8]

Sergeant Taylor Peirce, 22nd Iowa Infantry, writing to his wife on Jan. 29, 1865, at Savannah, Georgia: "I have a wall tent all to myself. I have a good floor in it and a frame work up in the inside to keepe it square and nice and yesterday I went to work and built me a nice little brick fire place and chimney and put up a pair of bedsteads and got some of the long moss and have just as good a bed as I wish. You had better think I am enjoying myself to-day."[9]

✎

At least one enterprising southerner tried to capitalize on the value of Spanish moss.

Gideon Lincecum, sixty-eight-year-old naturalist, in a letter to a friend from Washington Co., Texas, on Oct. 27, 1861: "People from hereabouts are all gone to the war; and the soldiers having appropriated all my blankets, I have succeeded in inventing a machine to spin the long moss. The intention is to supply my bed with a substitute from my absent blankets, and I am cleaning and spinning it every day. Lincoln will not be able to freeze me into subjection, so long as the gray *Tillandsia usneoides* floats to the breezes beneath the branches of our venerable forest trees."[10]

Gideon Lincecum, sixty-eight-year-old naturalist, in a letter to Confederate Secretary of the Treasury C. G. Memminger from Washington Co., Texas, on Dec. 14, 1861: "I constructed a machine to spin the long moss (*Tillan[d]sia usneoides*) of our climate. Also, a loom to weave it, and have succeeded in producing a blanket composed entirely of the cleaned moss, that is valued by those who have seen it, at $20. It is indeed a superior article, weighing 18 pounds. Carpets, rugs, bed blankets, saddle blankets, and if Kentucky had remained out of the confederacy, cotton bagging

of a superior quality, could be supplied from this abundant natural production. It is quite strong and very durable."[11]

Others found Spanish moss such a curiosity that they sent it home to
family and friends.

**Captain Jacob Ritner, 25th Iowa Infantry, in a letter to his wife from
Young's Point, Louisiana, on February 22, 1863:** "I will send you a specimen of moss that grows on the branches of trees down here. It hangs down
all over the trees, sometimes two or three yards or more long and looks
very odd—it makes a good bed to sleep on."[12]

**John M. Follett, 33rd Illinois Volunteer Infantry, at Milliken's Bend,
Louisiana, in a letter to his sister on April 5, 1863:** "Enclosed you will find
some moss such as hangs on every tree in the woods in these parts. It is for
you."[13]

**Private Isaac Jackson, 17th Ohio Battery, at New Orleans, Louisiana, on
May 17, 1864:** "That Moss I sent home was of a bright green color when I
pulled it. It may change its color. It is the same as the moss that they use up
there to make carriage cushions."[14]

SUMAC

Sumacs (*Rhus* spp.) are shade-intolerant shrubs or small trees usually having compound leaves. About eight species are found in the eastern United
States. Staghorn sumac (*R. typhina*) is named for its appearance in winter
when bare branches stand out in the shape of a stag's antlers. Poison sumac
(*R. vernix*), like the closely related poison ivy (*R. radicans*), contains the
allergen urushiol that causes severe dermatitis in many people. Although
not preferred wildlife plants, sumacs can be important to birds such as
quail during severe winters.

Sumac was important during the nineteenth century as a raw material
in the tanning industry. Sumac mills were common in the East and worked
on the same principle as gristmills. During the Civil War, millstones that
were later turned by engines were powered by water or horses and mules.
One source reported that a horse and three mules were used to convert

150 tons of sumac leaves into 130 tons of sumac "sauce."[1] A trade magazine stated that bark tanning of leather involved a twelve-hour sumac liquor bath.[2] Tannin from sumac was sought when light-colored, supple leathers, such as those used in bookbinding, were required. An unknown amount of nonnative sumac was imported by the Union to supplement that gathered locally for the industry.

Sumac was also an important source of bright red and black dyes during the Civil War both for the commercial mills of the North and the southern housewife's dye pots. Smooth sumac (*R. glabra*) was used in shoe wax, and the drupes were fermented into vinegar. "Sumac-ade," a refreshing drink, was made from at least two species. Sections of the pithy, easily hollowed stems were inserted into sugar maple bore holes to collect sap into pans below. Medically, the various species of sumac were used in the treatment of hemorrhoids, ringworms, syphilis, gonorrhea, rheumatism, and "putrid fevers."[3]

Poison sumac was at the center of one Civil War story that ended with an odd twist: "Private William McKesson Blalock resided in Caldwell County with his young wife of 20, Sarah Malinda Pritchard Blalock. On March 20, 1862, Sarah, being a truly loving wife, donned male attire, and enlisted, as 'Samuel' Blalock, with her husband in company F, of Zebulon Vance's famed 26th Regiment North Carolina Troops. A month later William Blalock was discharged for hernia and 'poison from sumac,' at which time Samuel (Sarah) immediately confessed her sex, and was ultimately discharged with her husband."[4] Private William Blalock is alleged to have intentionally wallowed in poison sumac for the purpose of obtaining a medical discharge.

Rev. Francis Springer, chaplain, 10th Illinois Cavalry, describing the debriefing of a spy at Rhea's Mill, Arkansas, in December 1862: "While the general [Blunt] listened & asked questions, he [the spy] plied his pocket knife most industriously on stick after stick that he cut from the lowly sumac that had grown on the spot."[5]

SWEETGUM

Distinctive star-shaped leaves identify sweetgum (*Liquidambar styraciflua*) trees. Found throughout the eastern half of the United States, sweetgum grows to 150 feet tall and occurs most abundantly on rich alluvial soils of the southeast. Underappreciated for its wildlife values, sweetgum is important to several species of migrating spring warblers, each of which uses different parts of the tree to forage for insects.[1] The lustrous heartwood of sweetgum, known as red gum in the lumber industry, was exploited in the twentieth century for such products as Singer sewing machine cabinets and Chevrolet car bodies. During the Civil War, Confederate surgeon general Samuel P. Moore officially directed all of his medical officers to make available indigenous astringents including sweetgum for the treatment of bowel complaints among sick soldiers.[2] Soldiers of both sides sought the plant for curative purposes.

Captain Charles B. Haydon, 2nd Michigan Infantry, near Vicksburg, Mississippi, on June 30, 1863: "The weather is terrible hot & I have felt very much wilted all day. . . . I peeled a quantity of sweet gum bark & chewed it to regulate my bowels. I ate nothing & resolved to be well tomorrow."[3]

Physician Esther Hill Hawks at Jacksonville, Florida, on Feb. 25, 1865: "Capt. Gates brought me a cane, from a sweet gum tree, with which I hope to be able to walk, before many days."[4]

Sweetgum bark mixed with that of maple and copperas produced a purple dye,[5] and the fruits were used in a unique type of lighting in the South. "In the absence of any of the ordinary materials for lighting, the good woman of the house had gone to the woods and gathered a basketful of round globes of the sweet gum tree. She had taken two shallow bowls and put some lard, melted, into them, then placed two or three of the sweetgum balls in each of the vessels, which, soon becoming thoroughly saturated with the melted lard, gave a fairylike light, floating round in the shallow vessels of oil like stars."[6] *Liquidambar,* the genus of the sweetgum tree, translates as "liquid amber" and refers to the waxy sap, which was often chewed like chewing gum.

Confederate sympathizer Myra Inman at Cleveland, Tennessee, on Jan. 8, 1863: "We went after gum wax this eve. . . . Cousin Mary Harrison and I, also Matt Gentry, went over to a gum wax tree, did not get any. Cousin Juliet and Matt Fox went to another gum tree, got a good deal."[7]

As a ruse to deceive an opposing force of the true strength or even presence of an army, logs were sometimes disguised as cannons and called "Quaker guns."

Private Robert A. Moore, 17th Mississippi Regiment, near Leesburg, Virginia, on Dec. 29, 1861: "Some of the boys have been over near the battleground discovered two gum logs mounted on our breastworks as cannon."[8]

However, in at least one instance sweetgum logs were actually used as barrels of functioning cannons. The "Sweet Gum Battery" comprised six 6-pounders and one 12-pounder and was manned by the 33rd Missouri Volunteers at Spanish Fort, Alabama, in early 1865. A description cites: "They were made of sweet gum wood, and banded at the muzzle and breech with a band of iron about one inch wide and one-quarter of an inch thick. The gun and carriage were separate, the carriage being a block of wood with a socket for the breech of the gun, giving the gun an elevation of about 45 degrees. The ordinary 6 and 12-pound shells were used, the surface being coated with turpentine to secure ignition of the fuse . . . . The men became so expert as to be able to burst a shell within the size of an army blanket at 500 to 600 yards distance."[9]

    Sweet Gum Stable was a stop of the Underground Railroad located at the corner of Main and West Seventh Streets in New Albany, Indiana. Fugitive slaves were harbored there after crossing the Ohio River on their way to freedom.[10]

SYCAMORE

Sycamore (*Platanus occidentalis*), often called "plane tree" during the Civil War era, is one of the largest hardwoods in eastern North America, growing to heights of 170 feet with a diameter of ten feet. It is easily recognized by its large lobed leaves and light-colored bark that flakes off in long thin sheets. Sycamore grows best in rich, moist, riparian soils where the hollows in older trees provide homes for wood ducks, owls, and squirrels. During the Civil War sections of sycamore logs were a favored material for butcher blocks. The wood was also used for ox yokes, crates, tobacco boxes, cross-ties, and cooperage.[1] In the South, the globe-shaped fruits, like those of sweetgum, were used for lighting when soaked in oil. Referring to a Catholic nun working at a Confederate hospital in Shelby Springs, Alabama, a writer penned, "she made her profession on a makeshift altar on a cold January morning in 1863 with sycamore balls burning in lard oil for lights instead of candles."[2]

Private Theodore F. Upson, 100th Indiana Infantry Volunteers, on Nov. 24, 1864, after the battle of Griswoldville, Georgia: "We had no coffins, but I could not bear to think of putting my old friend into his grave in that way. I remembered that at a house a short distance away I had seen a gum or hollow sycamore log of about the right length and size. We got it, split it in halves, put one in the grave dug in the sandy soil, put his lifeless body in it, covered it with the other half, filled up the grave and by the light of a fire we had built with the rails, marked with a peice of lumber pencil his name, Company, and Regiment."[3]

Private Wilbur Fisk, 2nd Vermont Volunteers, near Warrenton, Virginia, in a letter to his hometown newspaper on Aug. 13, 1863: "We had been obliged to go nearly three miles to get this one luxury [foraged milk], to cross two streams, one on loose flood wood and the other on the trunk of a huge sycamore tree, where the torrent dashed and foamed as if it would delight to engulf us for ever milk and all."[4]

Rev. Francis Springer, chaplain, 10th Illinois Cavalry, in northwestern Arkansas, on Aug. 16, 1863: "we camp at the Big Spring on Flint Creek. . . . This fine large spring of clean cool water with the surrounding stand of oaks, sycamore, & persimmons, is truly grateful to weary soldiers."[5]

WALNUT, HICKORY, AND PECAN

Walnuts (*Juglans* spp.), hickories (*Carya* spp.), and pecans (*Carya* spp.) are deciduous trees in the same taxonomic family. Dense, hard wood and edible nuts are characteristics of most species. Two types of walnut (black walnut, *J. nigra*, and butternut, *J. cinerea*) and about twenty species of hickories, which include pecans, are found in the eastern United States. Most are large trees with alternate, compound leaves. Members of this family grow in a variety of habitats from swamps to mountaintops. The large, fleshy fruits of many species are relished by humans and a host of wildlife species including squirrels and other rodents, many types of birds, rabbits, raccoons, fox, deer, and bear.

During the Civil War, large quantities of black walnut lumber were manufactured into gunstocks, a trend that continued into the world wars of the following century. Easy to work and resistant to splitting, the wood was an ideal component of the standard shoulder weapons of the day. The only drawback was the lengthy period of time required to properly dry walnut lumber. At a convention of southern gunsmiths in Atlanta in August 1861, one participant reported, "the greatest difficulty was to get wood for the stocks; that wood of one or two years was not sufficiently seasoned. It ought to be cut twenty years."[1] Obviously the urgency of the war precluded following this advice. The same general properties of black walnut were valued in its use in furniture, cabinetry, and musical instruments. Coffins, railroad crossties, posts, and parts of ships were made of walnut because of the wood's durability. As a medicament, Confederate surgeon Francis Porcher reported that black walnut concoctions were used to cure ringworms, toothache, and sores. He also stated, "Walnut leaves soaked in water for some hours, then boiled and applied to the skins of horses and other animals, will prevent their being bitten or worried by flies."[2] Especially in the Deep South, green walnut hulls were crushed and added to stagnant water bodies where they acted as a stupefying substance on fish that were then gathered to eat. An uncommon characteristic of black walnut stems is the presence of chambered pith. This quality made the species a favorite plant of pipe-smoking soldiers as the twigs were easily hollowed into pipe stems.

Butternut, also known as white walnut, is smaller than black walnut and more restricted in range. Its uses during the Civil War were similar to those of black walnut and included building lumber, fence posts and rails, cattle troughs, wooden dishes, and coach paneling. The term *butternut* is best known as a moniker for Confederate soldiers who wore uniforms died a yellowish-brown color using butternut hulls. Black walnut hulls produced gray to black dyes. The nuts of both species provided an important, seasonal dietary supplement to many during the Civil War.

Private John M. King of the 92nd Illinois writes in October 1863 at Harrison's Landing, Tennessee, of the Confederates just across the river: "Their uniforms consisted of broad-brimmed black hats and their clothes had two shades of color. It was all, or nearly all, home-made, butternut and sheep's gray were the colors. The butternut was made from the bark of butternut and walnut trees and was usually cotton goods."[3]

John B. Jones, clerk to the Confederate secretary of war, in Richmond, Virginia, on May 10, 1863: "Detachments of Federal troops are now marching into the city every few hours, guarded by (mostly) South Carolinians, dressed in home-spun, died yellow with the bark of the butternut-tree."[4]

William L. Nugent, 28th Mississippi Cavalry, near Vicksburg, Mississippi, in a letter to his wife on June 22, 1862: "We have moved our Camp to a cool shady walnut grove, where we have an abundance of good cistern water to drink & enough to eat of its kind."[5]

Private Melville Follett, 42nd Illinois Infantry, near Stones River, Tennessee, on Feb. 8, 1863: "In the afternoon Elliot, Bennett, Goodrich, Stebbins and myself went across the river and got some walnuts. Drank sugar water and had a good time generally."[6]

Confederate sympathizer Myra Inman at Cleveland, Tennessee, on Dec. 17, 1862: "Cousin Han brought us in some nice walnuts."[7]

Confederate sympathizer Myra Inman at Cleveland, Tennessee, on Oct. 26, 1863: "I ate a walnut last night and it made me sick, have been in bed all day."[8]

Sergeant Charles B. Haydon, 2nd Michigan Infantry, near Washington, D.C., on Aug. 7, 1861: "This is a rich country in wild fruits & nuts. I never saw trees loaded as they are here with butternuts, black walnuts, beechnuts, chestnuts, persimmons, hazelnuts, to say nothing of blackberries, grapes &c."[9]

Private Robert M. Holmes, 24th Mississippi Volunteers, near Shelbyville, Tennessee, on Jan. 31, 1863: "we built fires & remained by them until after sonrise . . . & not having anything to eat we geathered up some walnuts that was near by & made our breackfast on them."[10]

Assistant Surgeon William Child, 5th New Hampshire Volunteers, in a letter to his wife from near Sharpsburg, Maryland, on Oct. 22, 1862: "The leaves are yellowing and in the woods around the ground is covered with walnuts and acorns. I think I will get a man to collect me a bushel of the small walnuts and send them home in a box by express, but don't know how it may be."[11]

Corporal Edmund D. Patterson, 9th Alabama Infantry, at Manassas, Virginia, on July 31, 1861: "Today with heavy hearts we buried our friend and comrade. He sleeps his last long sleep beneath the shady boughs of a walnut tree, on the green grassy hillside, another martyr to truth."[12]

During the Civil War smoke of green hickory wood of various species wafted from rural smokehouses North and South as a favored fixative to cure and flavor meats in an age before electric refrigeration. The strength and hardness of hickory wood made it a preferred choice for the heads of cannon ramrods, tool handles, agricultural implements, wagon-wheel spokes and hubs, yokes, cogs of mill wheels, bookbinder presses, knitting needles, bands for cotton bales, ship fittings, and hoops for barrels and boxes. Several years before the Civil War small hickories used in hoop making were already becoming scarce in some areas. "Of the numerous trees of North America east of the Alleghany mountains, none except the hickory is perfectly adapted to the making of hoops for casks and boxes. For this purpose vast quantities of it are consumed at home, and exported. . . . When it is considered how large a part of the productions of the United States is packed for exportation in barrels, an estimate may be formed of the necessary consumption of hoops. In consequence of it, young trees proper for this object have become scarce in all parts of the country which have been long settled."[13] Hickory wood differs from walnut in being subject to rot and insect damage when in contact with the ground. The dense wood when burned resulted in high-quality charcoal and ashes containing choice soap-making lye. Hickory bark boiled with alum was used to produce a yellow dye. Southerners more than others made snuff dip sticks by chewing the end of hickory twigs into a brush.

Private Van R. Willard, 3rd Wisconsin Volunteers, in Middle Tennessee during the autumn of 1863: "The [snuff] dip cup and dip stick or brush are indispensable articles of furniture in almost every wealthy family in Tennessee. . . . The dip stick is usually made from a round bit of hickory, chewed at one end so as to make a kind of brush of it. This they dip into the cup and then put it in their mouths, chewing and rolling it around as men do cigars. They go about the house with their dip sticks in their mouths."[14]

Sergeant Edwin H. Fay, Minden [Louisiana] Rangers, near Guntown, Mississippi, in a letter to his wife on Aug. 13, 1862: "I sat up nearly all night and we started by daylight the next morning and came two miles above Guntown where we encamped in a woods lot, a hickory grove, one of the most pleasant places I ever saw if it was not for the dust which is oppressive and it is right on the Roadside."[15]

Refugee Sarah Morgan Dawson near Port Hudson, Louisiana, on Sept. 24, 1862: "In this case the Confederate carriage is a Jersey wagon with four seats, a top of hickory slats covered with leather, and the whole drawn by mules."[16]

Private Robert M. Holmes, 24th Mississippi Volunteers, near Shelbyville, Tennessee, on Nov. 28, 1862: "It is very cloudy and cool today. Most of the men are out hunting scaly boxes [hickory nuts] to eat."[17]

Private Robert M. Holmes, 24th Mississippi Volunteers, near Eagleville, Tennessee, on Dec. 10, 1862: "We havent much to do to day and most of the men are out in the woods hickrynut hunting and about country houses trying to bye something to eat allthough they get as much as they should eat in camps they desire something better."[18]

Reuben A. Pierson, 9th Louisiana Infantry, in a letter to his sister from near Centerville, Virginia, on Aug. 5, 1861: "The growth is mostly hickory but there are other kinds of timber and a variety of it around here, some fine poplar groves in the lowlands and plenty of shrubbery in many places."[19]

Confederate sympathizer Myra Inman at Cleveland, Tennessee, on Dec. 24, 1863: "Rhoda and I fixed up a few ground nuts, walnuts and hickory nuts for Stephney's [Christmas] stocking. Oh, so sad is our life at this time."[20]

Sergeant Alexander Downing, 11th Iowa Infantry, at Corinth, Mississippi, on Oct. 28, 1862: "The ramparts are faced on the outside with long woven baskets of hickory withes and filled with earth to keep them in shape."[21]

The name *pecan* is of Native American origin and was used to describe nuts that required a stone to crack. Pecans (*C. illinoensis*) are a type of hickory and grow naturally along the river bottoms of eastern North America and south into Mexico. Old, wild trees can exceed one hundred feet in height and three feet in diameter. The well-known fruit of pecan trees was an important food for humans and wildlife for thousands of years before Europeans arrived. Wild pecans exhibit great variety in nut size, shape, thickness of shell, and ripening date. Within this diversity an occasional native tree was discovered with unusually large, thin-shelled, sweet nuts. In 1846, a Louisiana slave successfully grafted one of these superior wild pecans onto a typical stock. After the Civil War the freed slave, Antoine, went on to be honored at the Philadelphia Centennial Exposition for his clones that became the first official plantings of improved pecans.[22] The uses of pecan during the Civil War were similar to other hickories.

Gideon Lincecum, seventy-year-old naturalist, in a letter to a friend, from Washington Co., Texas, on March 26, 1863: "The pecan bark, treated in the same manner, as directed for the black oak bark, will produce equally fine yellow colors [for dying fabric]."[23]

Private George A. Remley, 22nd Iowa Volunteers, in a letter to his father from New Iberia, Louisiana, on Oct. 8, 1863: "I have seen some persimmons and walnuts but they are not very plenty. Pecan nuts are abundant. They are very good when perfectly ripe and taste much like hickory nuts but when green they taste bitter like the white-hickory nuts you have in the yard."[24]

Kate Stone, Brokenburn Plantation near Milliken's Bend, Louisiana, on Oct. 22, 1861: "The boys have been thrashing and cutting pecan trees and have brought in lots of nuts. We hate for them to cut the trees. Shall stop it."[25]

Private Isaac Jackson, 17th Ohio Battery, at Opelousas Prairie, Louisiana, on Oct. 20, 1863: "We gathered oranges as we went along by the sacks full, and pecan nuts as plenty as walnuts are at home."[26]

WILLOW

Worldwide, several hundred species of willows (*Salix* spp.) grow most often on wet sites in the northern hemisphere. At least four species are found in the Civil War arena. Willows frequent the banks of streams and rivers, helping to prevent shoreline erosion while simultaneously providing food and shelter for beavers, rabbits, and deer. Bees make copious amounts of honey from willow nectar, and some species of butterflies and moths feed exclusively on willow leaves.

The manufacture and marketing of baskets and other wicker ware from the young pliable stems of willow was once a moderate-sized industry in the United States. Products made from European species of willow were often preferred over native willows, resulting in an import trade. In the six years prior to the Civil War, $815,847 worth of basket ware and $236,842 worth of willow raw materials were imported, mainly through New York. Most commercial basket production in the United States was in the Northeast.[1] Native willow products such as baskets and fish traps were undoubtedly made in the South, but a wartime embargo forced an increase in local production, as described by an Alabama woman: "Willow wickerwork came in as a new industry with us. We learned to weave willow twigs into baskets of many shapes and sizes. . . . The switches were gathered when the willows were flowering, and stripped of bark and leaves; what was not wanted for immediate use was put by in bundles, to be used in our leisure hours. When placed in warm water the withes were soon as flexible as if freshly gathered and peeled, and were as easily woven into varied kinds of wickerwork." She also reported making dye from willow bark.[2] Willow lumber is soft and light, but not durable, and was used to make boxes, toys, artificial limbs, cheap furniture, and boats. The medicinal values of willows have been recognized for centuries. Its bark contains salicylic acid, a precursor to aspirin, and was thought during the Civil War to yield an inferior substitute for quinine. Ground charcoal made from willow was considered a topical antiseptic, a laxative when taken in doses of ten to fifteen grains, and a prophylactic for yellow fever. The characteristics of willow charcoal, however, were most valued as a component of gunpowder. Georgia newspapers ran the following advertisement in 1862: "To Contractors—Willow wood wanted—Five hundred cords willow will be contracted for, to be delivered on the line of the canal, at the government powder factory, at Augusta, Ga., at the rate of not less than one hundred and fifty cords per month, commencing the 1st of December next. The willow may be of any size, the smaller branches being preferred; the larger sticks must be split into parts not larger than the arm. It must be cut into uniform lengths of three feet, and each cord will measure fourteen feet long, three feet high, and three feet broad, containing one hundred and twenty-six cubic feet. The bark must be carefully peeled of at the time of cutting."[3]

Submerged or overhanging willows could prove troublesome for stream navigation, piercing hulls and knocking down smokestacks. During the Yazoo River expedition, Union Admiral Porter wrote that his flagship *Cincinnati* ran into a six-hundred-yard bed of willows under a full head of steam, "and there she stuck; the willow wythes caught in the rough iron of her overhang, and held her as if in a vise." Taking advantage of the situation, Confederates pounded the flotilla with an artillery cross fire, and only with the greatest efforts was the *Cincinnati* freed to make her escape and end the expedition in failure.[4]

Private Isaac Jackson, 83rd Ohio Volunteer Infantry, in front of Vicksburg on Jan. 30, 1863, describing recent action: "The place where the men crossed, there was a sand bar and a row of willows in the middle of the bayou. . . . The men would run through a shower of bullets, fall down behind the willows and then get up, run across and get behind the levee."[5]

Private Van R. Willard, 3rd Wisconsin Volunteers, near Chickamauga Creek, Georgia, in May 1864: "The Chickamauga is not, as many suppose, a dark and sluggish stream, but a clear, sparkling creek over hung with willows, with green mossy banks where the pale lilies and watercress dip their beautiful heads in the passing wave."[6]

Captain William J. Seymour, 1st Louisiana Brigade, describing Willow Pump, Virginia, on Sept. 30, 1864: "This little settlement takes it's name from the fact that on the side of the pike at this point, a steady & abundant stream of pure, cool water gushes forth from a hole in the trunk of a Willow tree, coming up from a spring concealed from view under the roots of the tree."[7]

Grapevine bridge built May 27–28, 1862, over the Chickahominy River, Virginia, by the 5th New Hampshire Infantry. Construction of roads and bridges consumed countless trees.

Library of Congress, Prints and Photographs Division

Brush arbor at Petersburg, Virginia, summer quarters camp. Plants were often used to provide shelter from harsh weather.

Library of Congress, Prints and Photographs Division

Chevaux-de-frise in front of Fort Sedgwick, Petersburg. Plant materials were used in most implements of war, from cannon carriages to gabions.
Library of Congress, Prints and Photographs Division

Confederate palisades at Atlanta. Construction of fortifications often resulted in large-scale elimination of vegetation to maintain open fields of fire.
Library of Congress, Prints and Photographs Division

Palmetto logs used to reinforce the channel side of Fort Sumter, 1865.
Library of Congress, Prints and Photographs Division

Sutler's bomb-proof "Fruit and Oyster House," Petersburg. Soldiers welcomed every opportunity to procure fresh seafood.

Library of Congress, Prints and Photographs Division

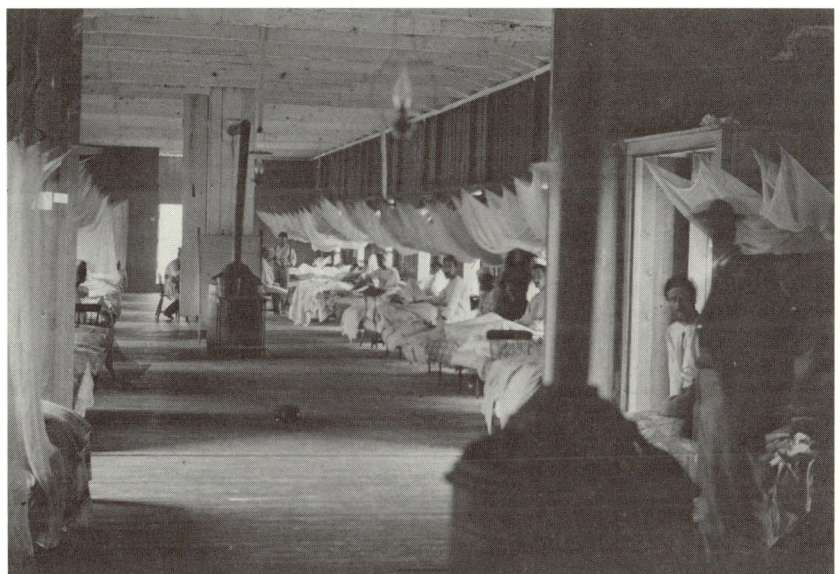

Harewood Hospital, Washington, D.C., with mosquito nets over beds. The netting was provided for comfort, as the link between mosquitoes and malaria was not discovered until after the Civil War.

Library of Congress, Prints and Photographs Division

Confederate fort on heights of Centreville, Virginia, with "Quaker guns," March 1862. Logs were sometimes disguised as cannon in ruses to misrepresent the strength of a position to the enemy.

Library of Congress, Prints and Photographs Division

Men repairing railroad after Battle of Stone's River near Murfreesboro, Tennessee, 1863. Millions of trees in the form of ties, car siding, and fuel for locomotives were used by railroads during the Civil War.

Library of Congress, Prints and Photographs Division

Federal earthworks near Point of Rocks, Bermuda Hundred, Virginia, 1864. Large-scale fortifications and earthworks changed the ecology of an area for years, usually by setting back plant and animal successional stages.

Library of Congress, Prints and Photographs Division

PART II

FAUNA

They may fight till the buzzards are gorged with their spoil,
Till the harvest grows black as it rots in the soil,
Till the wolves and the catamounts troop from their caves,
And the shark tracks the pirate, the lord of the waves.

—OLIVER WENDELL HOLMES, SR.
  "Brother Jonathan's Lament for Sister Caroline"
  March 25, 1861

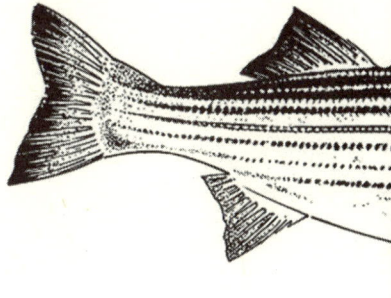

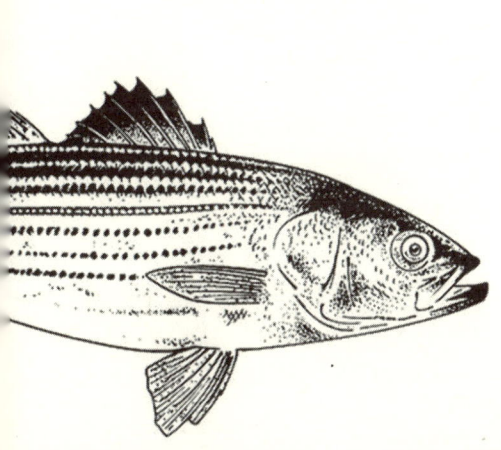

INTRODUCTION

Fauna is an assemblage of animals of a defined region or time. Like plants, the classification of animals continues to change as new scientific techniques to determine kinship, such as DNA analysis, are developed. Animals have traditionally been classified into tiers based on physical similarities such as the presence or absence of backbones, shape of teeth, or whether or not the young are nursed. Invertebrates, those lacking backbones, include among others insects, mollusks, and various types of worms, and comprise millions of species. Vertebrates, animals with backbones, are less numerous but contain the well-known mammals, birds, fish, reptiles, and amphibians. A recent survey of the native, terrestrial (land-dwelling), vertebrate fauna in most of the area covered by this book yielded 575 species. Of these, 102 (18 percent) are mammals, 112 (19 percent) are reptiles, 130 (23 percent) are amphibians, and 231 (40 percent) are birds.[1] Almost 500 species of freshwater fishes are found in the same region.[2] The fauna represented in this book had practical value to humans and/or were mentioned in diaries, journals, and letters during the Civil War.

Animals are referred to in anecdotes in many of the same contexts as plants, including food and habitat components.

Private Isaac Jackson, 17th Ohio Battery, at Berwick City, Louisiana, on Oct. 6, 1863: "As far as we have gone, we have seen a very fine country. Oysters are plenty here at 10¢ per Doz. And Crabs, we can catch plenty of them right here in the bay. I have caught plenty of them. We just tie a piece of meat on a string and throw in. They catch hold, and we haul him to the top of the water when we dip him with a net fixed for the purpose. Hundreds are fishing for them every day. They are excellent eating. The meat is very white and sweet, a little troublesome getting out of the shell."[3]

Corporal Rufus Kinsley, 8th Vermont Regiment, describing Ship Island, Mississippi, on April 7, 1862: "East end covered with pine, and various other trees; and inhabited by alligators, and all the snakes in the catalogue; besides wild fowl in abundance. A great variety of flowers. West end, all sand."[4]

Wild animals were considered objects of incessant curiosity. In an era before conservation this often parlayed into attitudes that promoted wanton waste:

Private Nelson Stauffer, 63rd Illinois Infantry, at Washington, D.C., on May 31, 1865: "Went to the city, and inspected the Smithsonian Institute. Saw birds and animals of every description and from almost every nation."[5]

John Hay, assistant secretary to President Lincoln, near Beaufort, South Carolina, on Jan. 25, 1864: "Shot a sea gull who fell in shallow water, couldnt get at him, ran aground. . . . I took the gun & after much maneuvering killed 2 crows at a shot."[6]

A common context unique to animals was their role as "vermin" in the form of chiggers, ticks, mosquitoes, and the like that dealt misery to all:

Corporal George M. Englis, 89th New York Volunteers, in a letter to his sister from Folley Island, South Carolina, on April 2, 1864: "I hadn't been in the tent only about 15 minuets before I had a general set to with the fleas. I think they came out bully though for I did not sleep mutch. Confound their eyes, if I had my way I would send them all to the Dry Tortugas for life, but you know it's all for the Union."[7]

Animals in this book are not listed in taxonomic order, as is often the case in works of pure natural history. Instead, the species or groups of similar species are arranged alphabetically by common name except for "Miscellaneous Invertebrates" and "Miscellaneous Mammals," which are at the end of this section.

ALLIGATOR

The name *alligator* is derived from the Spanish term *el lagarto*—the lizard—the label used by early Spanish explorers in Florida to describe the American alligator (*Alligator mississippiensis*). Alligators are found from North Carolina west to central Texas but rarely north of Arkansas and Tennessee because of their intolerance to lengthy periods of cold weather. Their required habitats include a variety of freshwater swamps, marshes, rivers, and ponds. Historically and currently, most alligators live in Louisiana and Florida. Once reduced to near extinction due to market hunting for their hides, the recovery of healthy alligator populations by wise management is an Endangered Species Act success story.

The American alligator is the largest reptile in North America. Males can reach thirteen feet in length and weigh six hundred pounds. The largest alligator on record measured nineteen feet two inches and lived in the marshes of southwest Louisiana.[1] Alligators are carnivorous and eat almost anything they can catch, including other alligators. A variety of insects, fish, turtles, birds, and mammals are common prey. Alligator attacks on humans are uncommon today and probably were the same during the Civil War period. Soldiers, however, did eat alligators on occasion.

Prior to the Civil War alligators were harvested for their skins, which were tanned into leather, and for the oil from their fat that lubricated cotton gins and steam engines. The war increased the demand for alligator leather for boots, shoes, and saddles for Confederate troops.[2] It is unlikely, however, that alligator products were important in the war except on a local basis.

Reflecting the general attitudes of the era, many Civil War participants considered alligators in the same category as bears, panthers, and wolves—vermin with no intrinsic beneficial values other than as targets for rifle practice. When mention of this species was made in their writings, this belief prevailed. Perhaps because many southerners were familiar with alligators, the species is more commonly mentioned in Union accounts, a result of its novelty. Just the sight of an alligator was worth noting in a letter or diary.

Private John Westervelt, 1st New York Volunteer Engineer Corps, near the mouth of the St. Johns River, Florida, on Feb. 27, 1864: "The banks of the river are mostly low and marshy. Occasionally we saw an alligator basking his horny hide in the sun."[3]

John Hay, assistant secretary to President Lincoln, near Fort Clinch, Florida, on April 25, 1863: "Tea with Col. Hawley & family. Best possible New England tea. A visit to the pet alligators. Bit in the thumb."[4]

Private Theodore F. Upson, 100th Indiana Infantry Volunteers, near Goldsboro, North Carolina, on March 27, 1865: "There are a great many small alligators and once in a while quite a large one in the pond above the mill. The boys have shot several. There was one that has kept well away but has been seen at times. I got on top of the mill to day and he showed up a long shot away. I raised the sights on my rifle and was fortunate enough to kill him. When the boys got him he measured 7 feet in length. The citizens and Darkies here think I am a wonderful shot."[5]

John S. Jackman, 9th Kentucky Infantry, aboard the steamboat *Waverly* on the Alabama River on Sept. 29, 1862: "The boys amused themselves by shooting at aligators lying out on the sand-bars and banks, sunning themselves."[6]

Corporal Rufus Kinsley, 8th Vermont Regiment, near Des Allemands, Louisiana, on June 3, 1862: "We killed four alligators on the way. I tried my rifle on two of them; put a ball in the right eye of each. One of them was thirteen feet long. We ate two of them for supper. Found the flesh, when boiled, more like a chicken's breast than any thing else."[7]

Surgeon's Steward C. Marion Dodson aboard the USS *Pocahontas* below New Orleans on May 10, 1864: "Alligators are quite abundent. One old fellow did not seem ill at ease when, fastened to an old log, we ran quite close to him."[8]

Private Galutia York, 114th New York Volunteer Infantry, in a letter to his brother on Jan. 7, 1863, from Fort Monroe, Virginia: "there was some of Co G men of 114 went out the other day and killed an alligator skinned him and brought his hide into camp they was on Ship Island [Miss.]."[9]

Private Galutia York, 114th New York Volunteer Infantry, in a letter to his parents on Feb. 1, 1863, from a ship near the mouth of the Mississippi River: "I saw an alligator day before yester day I should think he was 12 feet long they are thick down hear."[10]

Sergeant William D. Dixon, [Savannah] Republican Blues, near Savannah, Georgia, on Aug. 24, 1861: "We steamed off from Savannah in fine stile seeing nothing off of any importance. But the boys enjoyed the sport of shooting at alligators until we got opposite Thunderbolt where the ladies were out in numbers to look at us waving their handkerchiefs at us as far as they could see us."[11]

Major William J. Bolton, 51st Pennsylvania Volunteers, aboard the steamer *John H. Dickey* on the Yazoo River in Mississippi on June 16, 1863: "We also saw on our trip [upriver], many huge alligators."[12]

Corporal Rufus Kinsley, 8th Vermont Regiment, at Berwick Bay, Louisiana, on Feb. 23, 1863: "I enjoyed the ride hither very much. With the exception of two miles, the entire distance from Terrebonne to this place (28 miles) is an unbroken swamp, covered with a dense growth of large cypress trees; and from the top of the [train] cars where many of us stood, we saw hundreds of huge alligators, and large numbers of turtles, and a great variety of snakes, lying on large logs just above the surface of the water. We shot several, and shot at a great many. Woods vocal with birds."[13]

Lieutenant John Q. A. Campbell, 5th Iowa Infantry, near Richmond, Louisiana, on April 26, 1863: "On Holmes' plantation, we saw the 'first alligator.'"[14]

Captain William D. Dixon, [Savannah] Republican Blues, at Fort McAllister, Georgia, on May 3, 1863: "This morning there was some little excitement over an Alligator. It was playing around the battery all morning and the boys got [to] shooting at it but it did not seem to move him. I fired two shots at him, the second striking him in the jaw. He was brought in by the boat and revived, when the boys had some rare sport with it. He was seven feet long."[15]

British journalist William H. Russell near Charleston, South Carolina, on April 23, 1861: "Crossed the ferry—negro said river was full of alligators."[16]

Captain Charles B. Haydon, 2nd Michigan Infantry, near Vicksburg, Mississippi, on June 20, 1863: "There are some *alligators,* a good many snakes, lizards everywhere, plenty of mosquitoes, flies, bugs, tarantulas, horned frogs & other infernal machines too numerous to mention. I have not been far into the woods."[17]

Captain Charles B. Haydon, 2nd Michigan Infantry, near Vicksburg, Mississippi, on June 27, 1863: "The country is not so bad after all as I was at first led to believe. There are not so many snakes or other infernal machines as was represented. The alligators eat some soldiers [!] but if the soldiers would keep out of the river they would not be eaten."[18]

BATS

A diverse group with 1,200 species worldwide including 45 in the United States, bats are the only mammals capable of sustained flight. Most American bats are insectivorous and play important roles in controlling insect populations in many ecosystems. They navigate and find prey by using echolocation. A very small percentage of American bats are vectors for rabies. As a result bats often have undeserved, negative reputations and are even viewed as omens of bad luck, notions that were also common in the mid-nineteenth century.

Bats may have prolonged the Civil War. Gunpowder was made from a recipe of 75 percent potassium nitrate, 15 percent charcoal, and 10 percent sulfur. Traditional sources of potassium nitrate, also called "niter" or "saltpeter," included guano from seabird colonies usually on islands off the northeastern coast or other remote locales. For the blockaded Confederacy the procurement of niter was a critical issue. Cave-dwelling species of bats also produce large volumes of niter-rich guano over time. Accordingly, the Confederacy was able to maintain the war effort by exploiting virtually every known bat cave in the South. The Niter and Mining Bureau within the Confederate War Department was originally established for the purpose of procuring niter.[1] The process involved gathering the guano and placing it in large vats. Water that leached through the guano was collected and boiled down to yield potassium nitrate crystals, which were ground into a fine powder. Men who worked in the manufacture of the product were considered so valuable by the Confederacy that they were exempted from the draft.[2] Mexican free-tailed bats (*Tadarida brasiliensis*) in Texas caves forming colonies of more than a million individuals likely produced most of the guano for the South. Farther east other species, including the now endangered gray bat (*Myotis grisescens*), were important. Several hundred men, white and black, worked the limestone caves of northern Alabama. By September 1864 this area had produced 222,665 pounds of niter valued at nearly $238,000.[3] Such disruptive intrusions in active bat caves undoubtedly had harmful consequences for the resident bats. Without bats, however, the Confederacy would have run short of gunpowder long before it did.

John B. Jones, clerk to the Confederate secretary of war, in Richmond, Virginia, on Oct. 15, 1863: "I saw a common leatherwing bat flying over the War Department. What this portends I do not pretend to say, perhaps nothing. It may have been dislodged by the workmen building chimneys to the offices of the department."[4]

Assistant Surgeon Dr. Daniel M. Holt, 121st New York, near Harpers Ferry, West Virginia, on July 29, 1864: "Crossed the river on pontoon bridge and entered Harper's Ferry at 12m. Of all the miserable and forsaken holes I ever saw, this caps the climax. A perfect Arabia Petra.—A habitation for bats and owls. John Brown was a fool to make so much of the town as to ever set foot in it."[5]

BEARS

Black bears (*Ursus americanus*) were the largest wild, terrestrial animals found in that part of the United States where most Civil War campaigns occurred. Already reduced in range and numbers by hunting because of their omnivorous feeding habits that included depredation of livestock, gardens, and bee hives, bear populations thrived only in the larger, more remote forests, especially in southern swamps. Earlier, bears provided considerable food for settlers and Native Americans. Their fat, used for cooking oil, and skins were traded commodities throughout the colonial era. By the time of the Civil War, bear/human interactions usually resulted in the attempted extermination of the bear only because of perceived (but occasionally legitimate) competition with humans—a scenario common if any species of predator was involved. Then, as today, the presence of a bear created excitement on any occasion.

Kate Stone, Brokenburn Plantation near Milliken's Bend, Louisiana, on May 26, 1862: "In the afternoon there was a cry raised that there was a bear in the cane. The boys with their dogs and guns turned out in force ... as did all the Negroes who could get mules, while the others armed themselves with axes and sticks and cautiously approached the outskirts. The excitement ran high and we at the house had full benefit as it was in the canebrake just back of the yard. We could hear the barking of the dogs, the reports of the guns, and the cries and shouts of the whole party. It was very exhilarating. They returned in the highest state of excitement but without the bear."[1]

Rev. Francis Springer, chaplain, 10th Illinois Cavalry, at Fort Smith, Arkansas, on Dec. 31, 1863: "On another occasion I heard several citizens express surprize that the General [James G. Blunt] had become so much enraptured with a half-grown bear which some boys had on exhibition in town, that he procured the animal & was leading it through the streets on the Sabbath day to his quarters."[2]

Lieutenant Jacob Ritner, 25th Iowa Infantry, in a letter to his wife from near Delta, Mississippi, on Nov. 30, 1862: "Well, I have had my dinner. I had sassafras tea, bear-meat, and hard crackers. There are plenty of bears round here. A man who lives here killed two day before yesterday, and I got some of the meat."[3]

Private Isaac Jackson, 83rd Ohio Volunteer Infantry, at the mouth of the White River in Arkansas on Nov. 16, 1864: "This place we are camped on is an island formed by the Mississippi, White, & Arkansas Rivers and the Arkansas cut off which runs from the 'Ark' to the White River. We are not bothered the least by Rebs here and it said that there are plenty of bear and deer on the island. Several have been killed they say, but I have not seen any yet."[4]

The following passage, although out of the normal geographic scope of this book, may be a rare Civil War reference to the grizzly bear (*U. arctos horribilis*). Grizzly bears were still present in the Arizona Territory mountains at the time; however, black bears as a species can be black or any shade of brown.

Sergeant George O. Hand, 1st California Volunteers, at Tucson, Arizona Territory, on Nov. 10, 1862: "Some boys killed a large brown bear today. Mexicans & Indians will not eat it—superstition is the cause."[5]

Busby—a bearskin hat worn by some military units, especially those of German origin, in the Civil War.[6]

BIRDS

More than 700 kinds of birds are found in North America. At least 230 species live today in that part of the United States where most of the Civil War was fought. Five species—passenger pigeon, zenaida dove (*Zenaida aurita*), Carolina parakeet (*Conuropsis carolinensis*), Key West quail-dove (*Geotrygon chrysia*), and ivory-billed woodpecker (*Campephilus principalis*), which were present during the war, have since disappeared.[1] Others are much less common now. The diversity of bird life is great and often divided into groups such as waterfowl, birds of prey, shorebirds, songbirds, and so on. As with all animals, birds are not scattered randomly across the landscape but rather are adapted to various types of habitats. Some live in the forest interior, others on forest edges. Some are found in marshes, others only on the seashore. Many species live their entire lives in one area. Others participate in the mysterious phenomenon of migration, often flying thousands of miles seasonally between northern breeding grounds and southern wintering areas.

Civil War participants mentioned birds most often in the context of food. Birds were shot, trapped, poisoned, and procured by all available means. One method of the day involved the use of birdlime, a viscid, adhesive material spread on twigs and branches to entrap small birds. Birdlime was often made from the boiled, fermented bark of holly, but also could be concocted from mistletoe and elder.[2] No practical laws existed to protect birds during the Civil War, the first effective federal law coming a half-century later with the enactment of the Migratory Bird Treaty Act of 1918. Today, birds are separated into groups of "game" and "nongame" species. Game birds are those that can be hunted according to established regulations and include those that were traditionally pursued such as waterfowl, turkey, and quail. The remaining species are considered nongame birds and are protected year round. The distinction between game and nongame species during the Civil War was often blurry, especially for hungry soldiers.

Colonel Nathan W. Daniels, 2nd Louisiana Native Guard Volunteers [Union], Ship Island, Mississippi, March 12, 1863: "Had sea-gull cooked for supper and eat the same under the supposition that it was Duck, found it good, but a little tough."[3]

Private John Westervelt, 1st New York Volunteer Engineer Corps, at Morris Island, South Carolina, on Nov. 27, 1863: "A low sand bar that stretches a mile or so out to sea in front of our camp is covered with millions of curlems [curlews] a bird a little larger and verry much resembling the sea gull. I never saw so many of the feathered tribe together before. The bar is not only covered but the air over it is black with them. O how I wish I had a good shot gun. They are said to be an excellent bird to eat."[4]

Private Galutia York, 114th New York Volunteer Infantry, in a letter to his parents on Feb. 14, 1863, from a quarantine station below New

Orleans: "I wish I had the old gun down hear I would shoot black birds enough for one good stew they are first rate cooked."[5]

Captain Theodore A. Dodge, 101st New York Volunteers, near Stafford, Virginia, on March 21, 1863: "Cooper, the butcher, today went out shooting and brought in some nice fat robins, which he had deviled for supper. The Colonel has given him a shooting commission and he is to supply the mess with game, if he can. Some shot is to be got for him and he hopes to get a few snipe for us."[6]

During the Civil War, the northern bobwhite (*Colinus virginianus*) was commonly known as "partridge." This little quail thrived in the weedy fields and fencerows of farms and plantations throughout the Southeast. Still considered an epicurean delicacy, bobwhite numbers are greatly reduced today as a result of clean row-crop farming and pesticides. They were shot and captured using nets and baited, pyramid-shaped traps constructed of cane or small branches during the Civil War.

Kate Stone, Brokenburn Plantation near Milliken's Bend, Louisiana, on Jan. 17, 1862: "Warren sent up four partridges tonight. They were such sensible, happy looking little birds that I could not bear to have them killed and so turned them loose in the garden. He traps quite a number."[7]

Sergeant Edwin H. Fay, Minden [Louisiana] Rangers, at Okolona, Mississippi, in a letter to his wife on Feb. 22, 1863: "I want you to spin and make me a Partridge net by the time I come home."[8]

Sergeant Edwin H. Fay, CSA Engineer Bureau [formerly of the Minden Rangers], at Opelousas, Louisiana, in a letter to his wife on March 3, 1865: "Hope you have eaten your Sardines. Am glad Dr. Pattillo sent you some Partridges for I do hope that they will do you good."[9]

Gideon Lincecum, seventy-one-year-old naturalist, in a letter to a friend from Washington County, Texas, on Nov. 15, 1864: "Lysander used up the shot you left here killing partridges for the old Lady to eat during several spells of her ill health. I have on hand, however, 8 or 10 pounds of 4s which I will send to you the first favorable opportunity."[10]

Mary Boykin Chesnut at Richmond, Virginia, on January 28, 1864: "Bill of fare for my supper: Wild turkey, wild ducks, partridges, oysters, and a bowl of apple toddy made by Mrs. [Jefferson] Davis's recipe."[11]

A. L. Peel, Adjuntant, 19th Mississippi Regiment, near Fredericksburg, Virginia, on May 16, 1863: "Lt Bowen & McKie & I went hunting to kill birds for Maj Harkin who is sick. Killed one partridge. I shot several times with my pistol, missed."[12]

Confederate sympathizer Myra Inman at Cleveland, Tennessee, on Nov. 16, 1864: "Rained a little this eve. The boys went partridge driving this morn."[13]

⁊⦿

Wild turkeys (*Meleagris gallopavo*), historically common, suffered serious declines after the Civil War with the widespread logging of southern and eastern forests, only to rebound and become a model of successful conservation efforts today. As the largest game bird in North America (males commonly exceed twenty pounds), the wild turkey was a prized source of protein in the Civil War. Turkey-feather quill pens made Civil War correspondence possible for many.

John B. Jones, clerk to the Confederate secretary of war, in Richmond, Virginia, on March 19, 1864: "I saw a large turkey to-day in market (wild), for which $100 was demanded."[14]

Major David Pierson, 3rd Louisiana Infantry, in a letter to his father from the Yazoo River above Vicksburg, Mississippi, on April 3, 1863: "The woods, or I might rather say the canebrakes, abound with wild turkey, and now and then the boys waste a cartridge in killing one—which is strictly forbidden."[15]

Lieutenant Theodore A. Dodge, 101st New York Volunteers, near Harrison's Landing, Virginia, on July 3, 1862: "Such a country for productiveness I never saw. . . . Wild turkey and quail run through our camp all the time."[16]

Eliza W. Howland, Union nurse, in Washington, D.C., on Thanksgiving Day, 1861: "The dining room was the Sibley tent, charmingly ornamented with evergreens, and the dinner was a great victory in its way; for out of the little tent-kitchen appeared successively, oyster soup, roast turkey, cranberry sauce, canvas-back ducks, vegetables, and a genuine and delicious plum pudding that would do justice to any New England housekeeper."[17]

Refugee Kate Stone describing her travels through Louisiana's Tensas Swamp in April 1863: "The morning air was pleasantly cool, and as the red light crept up the sky we heard all kinds of wildwoods sounds—squirrels chattering in the trees, birds waking with a song, the calls of the wild ducks and turkeys, and three or four deer bounding into the woods just before us."[18]

Continental waterfowl populations during the Civil War are inestimable but numbered in the many millions. As the breeding grounds of most ducks and geese are north of the Civil War arena, they were usually encountered during migration or on the wintering grounds in vast bottomland hardwood swamps and coastal marshes.

Sergeant Edwin H. Fay, CSA Engineer Bureau [formerly of the Minden Rangers], at Opelousas, Louisiana, in a letter to his wife on Feb. 5, 1865: "There are the greatest abundance of wild ducks in this 'grand prairie' that I have ever seen. Negroes bring them in by horseback loads. There is a mulatto that a number of years ago when a good negro could be bought for $1000, brought $2200 and his master says that he gave him a gun and in two years he brought him back all the money killing ducks, geese and other game. I have seen him bring in 60 & 70 ducks and geese by 10 o'clock in the morning. He has an ox and a horse trained to feed along near the edge of a pond with which the prairie is filled & he walks along by its side till he gets 15 or twenty ducks huddled together (as they are not afraid of a horse (loose) or ox they will swim along as if nothing were near). He told me he would not fire at less than ten in a bunch and he sometimes kills 25 or 30 at a shot. These he brings to town and sells at $5.00 a piece in State money or 25 cents in specie."[19]

Confederate nurse Ada Bacot at Charlottesville, Virginia, on October 15, 1862: "I saw a flock of fourteen wild geese this afternoon flying south ward."[20]

Kate Stone, Brokenburn Plantation near Milliken's Bend, Louisiana, on Nov. 29, 1861: "The boys have been out hunting most of the day with poor success—one duck—but the woods are full of game and the lakes covered with ducks."[21]

Private Galutia York, 114th New York Volunteer Infantry, in a letter to his brother on Feb. 19, 1863, from a quarantine station below New Orleans: "wild geese are thick down hear but I haint got any gun to shoot them with alligators are thick."[22]

Lieutenant William D. Dixon, [Savannah] Republican Blues, Fort Jackson, Georgia, on Dec. 12, 1861: "I went ducking this morning. It is very cold."[23]

Lieutenant Colonel Charles B. Haydon, 2nd Michigan Infantry, recuperating at home in Decatur, Michigan, on Oct. 24, 1863: "I do considerable duck shooting, have good success."[24]

Lieutenant Colonel Charles F. Johnson, 81st Pennsylvania Volunteer Infantry, in a letter to his wife from Shippen Point, Virginia, on April 11, 1862: "The principle amusement of the men are gunning for wild Ducks, digging for Clams, raking for Oysters and fishing for trout; and like true soldiers, a few days of clear weather and a little recreation fully compensates for nearly two months privations."[25]

Major James T. Poe, 11th Arkansas Volunteers, at Reelfoot Lake, Tennessee, in November 1861: "The citizens informed us that there were many kinds of fowls there such as geese, ducks, swans, and in fact, all kinds of fowls known in that country and climate. Some of the boys desired to go to the lake to kill fowls. I did not object to them going so they kept it up from day to day as long as we remained on the main shore. The boys had good luck in fowling. One of them killed a large fine deer."[26]

Private S. O. Bereman, 4th Iowa Cavalry, near the Big Black River, Mississippi, on July 25, 1863: "We had a small skirmish today with a Staff Officer—Capt. Dayton. Our Sergt.—Housel—fired his revolver at a crane 'which was against the rule,' and the Capt. ordered us to tie him up to a tree by the thumbs which we positively refused to do. The Capt. got very wrathy and swore he would tie the whole of us up. We knew he could'nt do it by himself—but the 13th regulars would be here soon, & we knew they would be glad of a chance to tie us up, so two of the boys proceeded to tie him up, but he was released under arrest. I believe in discipline—but dont believe in being made a dog of by a petty staff officer."[27]

Bird vocalizations have long evoked emotions in humans. Whether the joyful mimicry of mockingbirds (*Mimus polyglottos*) or the lonely cry of whip-poor-wills (*Caprimulgus vociferus*), birdcalls were considered worthy of mention in many Civil War letters, journals, and diaries.

Major James A. Connolly, 123rd Illinois Infantry, writing to his wife on June 9, 1864, from camp near Acworth, Georgia: "Brass band are playing in very direction, and the mocking bird is making the leafy shade vocal with his attempts to imitate the brass music of 'Dixie,' 'Star Spangled Banner' &c &c."[28]

Mary Boykin Chesnut near Camden, South Carolina, on Dec. 6, 1861: "Here everything is fresh, bright, cool, sweet-scented; and a mocking bird is singing and a woodpecker at work—or a yellow hammer, for I cannot see the small bird which is making such a noise."[29]

Lieutenant John Q. A. Campbell, 5th Iowa Infantry, at Milliken's Bend, Louisiana, on April 20, 1863: "After a few day's stay in Louisiana, I am convinced that the inhabitants of this part of the State 'lived at home.' The green fields, the hedges, the mocking birds, and the many other adornments and songsters that Nature has lavishly bestowed on their land make it almost an Eden."[30]

Colonel Thomas W. Higginson, 1st South Carolina [African American] Volunteers, near Beaufort, South Carolina, on Dec. 20, 1862: "Mocking birds are abundant, but rarely sing; once or twice they have reminded me of the red thrush."[31]

Union Brigadier General Alpheus S. Williams in a letter to his daughter from Tullahoma, Tennessee, on Nov. 20, 1863: "We have very beautiful moonlights just now and an immensity of whippoorwills, and there are two mocking birds which begin their imitations every night in apple trees close to my tent. They mock everything from a frog to a crow. Some of their notes are beautifully sweet. The boys have tried to capture them but without success so far."[32]

Major James A. Connolly, 123rd Illinois Infantry, writing to his wife on May 15, 1864, from a camp near Resaca, Georgia: "It is now about nine o'clock at night, the moon is shining with a misty light through the battle smoke that is slowly settling down like a curtain, over these hills and valleys; the mournful notes of a whippoorwill, nearby, mingle in strange contrast with the exultant shouts of our soldiers—the answering yells of the rebels—the rattling fire of the skirmish line, and the occasional bursting of a shell."[33]

John S. Jackman, 9th Kentucky Infantry, near Dalton, Georgia, on April 21, 1864: "Clear to-night, the Moon shining brightly. I hear a whip-poor-will back of our cabin to-night, the first one I have heard this spring."[34]

Union Brigadier General Alpheus S. Williams describing the night before the beginning of the battle of Chancellorsville in a letter to his daughter from Stafford Court House, Virginia, on May 18, 1863: "The whippoorwills, which are thicker here than katydids up north, were whistling out their 'whip-poor-wills' as if there was nothing but peace on earth, and save the occasional crack of the rifle away off on the left there was a solemn stillness which was almost oppressive. Two immense hostile armies, over two hundred thousand armed men, lay within almost the sound of one's voice."[35]

Private Wilbur Fisk, 2nd Vermont Volunteers, near Richmond, Virginia, in a letter to his hometown newspaper on May 20, 1862: "The whipowill is singing merrily on a branch over my head and as we are to start again in the morning at precisely four o'clock I am reminded that it is time to seek repose."[36]

Colonel Thomas W. Higginson, 1st South Carolina [African American] Volunteers, at Port Royal Island, South Carolina, on April 12, 1863: "In the misty gray of the morning I rode out to the Ferry amid rose scents & the song of early birds—hearing for the first time the Chuck-Will's Widow, the Southern Whippoorwill, whose peculiar note I at once recognized."[37]

Private Van R. Willard, 3rd Wisconsin Volunteers, describing the Gettysburg battlefield in July 1863: "The night [July 2] was calm and still and starlit, not a breath of air stirring among the leaves, while the night birds kept up their songs—the whippoorwills gaily singing in the forests and the owls dismally hooting from some distant crag or blasted tree."[38]

John S. Jackman, 9th Kentucky Infantry, near Dalton, Georgia, on March 27, 1864: "I think the winter is now over. The spring-birds have set up their songs in the thickets; and this morning I heard the cooing of a dove—the first of the season."[39]

Confederate sympathizer Myra Inman at Cleveland, Tennessee, on Oct. 27, 1863: "Mother heard a snowbird today."[40]

Colonel Thomas W. Higginson, 1st South Carolina [African American] Volunteers, near Beaufort, South Carolina, on Dec. 5, 1862: "All the talk of the camp outside is fused into one cheerful & meaningless hum, pierced constantly by the wail of the hovering curlews."[41]

Sergeant George M. Englis, 89th New York Volunteers, in a letter to his mother from Point of Rocks, Virginia, on Feb. 22, 1865: "tis as pleasant here as June almost, & I have actually heard and seen bluebirds & robbins within the past week & today there is no need of fire to keep warm."[42]

Private John Westervelt, 1st New York Volunteer Engineer Corps, near Fort Monroe, Virginia, on May 16, 1864: "The flowers, trees and birds are the same as those of my own State . . . [New York]. One field of clover of many acres was completely filled with those saucy and amusing little songsters, the bobolinks."[43]

L. D. Bradley, Waul's Texas Legion, in a letter to his wife from Mud Island, Texas, on July 27, 1864: "it is, by all odds, the most desolate and out of the way place it has ever been my misfortune to be located in. There is not, nor ever has been, any house, or settlement of any kind, on it; not a tree, or even a bush as high as your head; nothing but sand and, for music, the continual roar of the surf, the whistling of the winds, & the melancholy cry, or rather croak, of the sea gulls, as they float lazily by, viewing with apparent, and, I think, just wonder our unusual presence."[44]

John Hay, assistant secretary to President Lincoln, at Key West, Florida, on March 7, 1864: "The seagulls that soar above the sea have their white breasts & inside wings splendidly stained with green by the reflection of the gleaming water."[45]

British journalist William H. Russell in Washington, D.C., on Nov. 27, 1861: "I had an engagement with Geoffroy which I did not keep to look at a whole lot of colibris or humming birds."[46]

Dr. Francis Bacon, Union physician, in a letter to his wife from Tybee Island, South Carolina, on Dec. 24, 1861: "The purple grey depths of the wood all flicker with scarlet grosbeaks [cardinals] like flames of fire, and quaint grey and brown northern birds flit in and out with the knowing air of traveled birds, and plan the nests they will build next summer, in spite of bombs and bayonets, in New England elms and alders."[47]

The survival of formerly abundant species such as the greater prairie chicken (*Tympanuchus cupido*) mentioned below is threatened today because of loss of habitat. Passenger pigeons, once the most common bird in North America with an estimated population of five billion, were already on the path to extinction during the Civil War because of commercial market hunting.[48]

Sergeant John Q. A. Campbell, 5th Iowa Infantry, in Boonville, Missouri, on Sunday, Jan. 5, 1862: "We have prairie chicken for breakfast and dinner. No chaplain—no preaching."[49]

Confederate sympathizer Myra Inman at Cleveland, Tennessee, on Feb. 5, 1863: "We saw five flocks of wild pigeons going northward this morn, a beautiful sight."[50]

Sergeant Lycurgus Remley, 22nd Iowa Volunteers, in a letter to his father from Rolla, Missouri, on Nov. 11, 1862: "There seems to be a good deal of game in this country: deer, wild turkeys, foxes, coons, rabbits, squirrels, quails, pigeons, etc."[51]

Large birds of prey are sometimes portrayed in bellicose images of strength and valor. America's national emblem, the bald eagle (*Haliaeetus leucocephalus*), was highly regarded metaphorically and otherwise during the Civil War. The most famous mascot of the war was a bald eagle known as Old Abe carried by Company C of the 8th Wisconsin Regiment. This bird allegedly witnessed numerous battles and skirmishes and survived the war to appear at fund-raising events for veterans and orphans.[52]

Article in Williamsport, Pennsylvania, *Lycoming Gazette* on April 24, 1861: "While the big flag was being raised over Langdon & Diven's Mill on Saturday, a large Eagle came and hovered for a few moments directly over the spot, and then departed due South. This incident is so well authenticated that there is no room for doubting its occurrence; and it is the more remarkable because an eagle is very rarely seen hereabouts. We have not seen one in fifteen years."[53]

Private John Westervelt, 1st New York Volunteer Engineer Corps, at Folly Island, South Carolina, on April 5, 1863: "There is an eagles nest in a pine tree over our camp."[54]

Private William D. Walton, 4th Pennsylvania Volunteers, in a letter to his sister from near Baltimore, Maryland, on Aug. 6, 1861: "May the wings of the American Eagle never be clipped."[55]

Birds that scavenge conjure up images in the mind's eye, most of which are morbid. Commonly called buzzards, two types of vultures are found in the United States. Turkey vultures (*Cathartes aura*) are the largest with a wingspan of six feet and a long, rounded tail. Black vultures (*Coragyps*

atratus) have a wingspan of four and a half feet and a short tail. Black vultures depend mostly on sight to locate carrion, while turkey vultures depend on a remarkable sense of smell. During the Civil War vultures played a critical role in disease prevention by dispensing of livestock carcasses and, infrequently, human casualties.

Sarah Wadley near Monroe, Louisiana (approximately sixty miles west of Vicksburg), June 5, 1863: "we hear that four Generals, Johnston and Bragg, and Rosencrans and Grant are concentrating large armies near Vicksburg, and we hear from the best and most direct sources that the Yankee dead lie in heaps about our entrenchments; it is horrible to relate, sickening to think, but so curious a fact that I must note it down, all the vultures have left this country, a carcass may lie for days untouched, those creatures have gone eastward in search of nobler game; how terrible is war!"[56]

Sergeant Alexander Downing, 11th Iowa Infantry, near Jackson, Mississippi, on July, 11, 1863: "We suffered for lack of water today, for the rebels in their retreat polluted the branches they crossed by killing and throwing into the streams their worn out horses and mules, hoping thus to strike a blow at us. Their march was marked by the buzzards flying above or feeding upon the carcasses of the slain animals."[57]

Private Richard H. Brooks, 51st Georgia Infantry, in a letter to his wife from near Charleston, South Carolina, on June 9, 1862: "There is more provisions thrown a way here than Blakely could destroy. The cold victuals an meat that we Let spoil is hauled of Everyday to the Buzzards."[58]

Colonel Thomas W. Higginson, 1st South Carolina [African American] Volunteers, on the St. John's River, Florida, on March 30, 1863: "James Rogers is a splendid shot & did the other day something I never saw done before. Shot a turkey buzzard on a bough & left it there, making ineffectual efforts to fly. He wanted one to stuff, but it did not fall, or not for a long time, though it was evidently shot. It would have been pathetic had the bird been more attractive; indeed it was rather so, as it was, to me."[59]

Assistant Surgeon Dr. Daniel M. Holt, 121st New York, at winter quarters on the Hazel River, Virginia, on Feb. 7, 1864: "Even crows, those most scary of all birds, devour putrid carcasses of mules, horses and camp offal with as little concern for man as if he were an inoffensive mouse."[60]

BODY LICE, TICKS, AND HARVEST MITES

Misery! The ordeals of Civil War soldiers were often aggravated by "vermin" as much as by the enemy. Body lice (*Pediculus humanus corporis*) vied with mosquitoes as the most dreaded of insect pests. One writer penned, "Like death, it was no respecter of persons. It preyed alike on the just and the unjust. It inserted its bill as confidingly into the body of the major-general as of the lowest private."[1]

Body lice are tiny parasitic insects that feed on human blood and live in the seams and creases of clothing. Their bite causes intense itching of the skin. Lice have three life stages. The eggs, called nits, may be attached to body hair and take about thirty days to hatch. Eggs hatch into nymphs, which resemble small adults and must have a blood meal to develop. Because of their color, adult lice were called graybacks. About the size of a sesame seed, adults also need to feed on humans to live.

Body lice are transmitted by direct contact with an infested person and through exposure to infected clothing or bedding. Soldiers frequently lived in an environment ideal for body lice transmission: poor hygiene and overcrowded living conditions. Infections persisted until clothes were boiled and bathing became routine, remedies not always available in wartime. Body lice can carry typhus and relapsing fever, although their part in these diseases during the Civil War is unclear.

Even if repugnant to their victims, lice sometimes provided recreation for bored soldiers, including prisoners. A number of lice each with its ardent supporters were placed in the center of a dish and the first to reach the edge was declared winner of the "louse race."[2]

Private John King, 40th Georgia Infantry, at Camp Chase Prison in Columbus, Ohio: "One can scarcely imagine that there could be any fun, any real amusement in watching the antics of a miserable louse, and certainly less to inspire a poetic thought. Robert Burns might write a poem on the creeping thing he saw on the ribbon of a pretty girl in a Kirk in Scotland, but is it possible that a poor forlorn prisoner in the destitution of the meanest poverty, could get a moment's real fun in playing with the miserable tormentor that had crawled over his back and rendered his life a constant affliction? What think you, gentle reader, of a louse race, of a regular pugilistic encounter between two champions of the genus pediculidae? I have witnessed both and have seen many a potato won and lost by the owners of these tormentors."[3]

Union nurse Hannah Ropes at a hospital in Washington, D.C., in a letter to her daughter, probably in December 1862: "You have no idea of a hospital, nor has anyone who simply calls in to see me. We get lousy! . . . My needle woman found nine body lice inside of her flannel waistcoat after mending the clothes that had been washed! And I caught two inside the binding of my drawers!"[4]

Captain Oliver Wendell Holmes, Jr., 20th Massachusetts Volunteers, in a letter to his father on June 13, 1862, from Camp Lincoln, Virginia: "Shall I confess a frightful fact? Many of the officers including your beloved son have discovered themselves to have been attacked by body lice."[5]

John M. Follett, 33rd Illinois Volunteer Infantry, at Vicksburg, Mississippi, in a letter to his wife on June 7, 1863: "Yesterday we drawed new clothes and then the whole regt went out and boiled all their old ones and killed all the lice, jiggers, fleas, ticks, ants, spiders, and every other thing with which infests our clothes. I drawed a new pair of pants, and as my old ones are tolerable good I will have two pairs, so I will boil them every week."[6]

Private Gottfried Rentschler, 6th Kentucky Volunteer Infantry [Union], in a letter to the *Louisville Anzeiger* from near Atlanta, Georgia, on July 25, 1864: "While crossing the creek General Hazen's Provost Marshal was severely wounded by a musket ball. . . . Four men had to carry him on the stretcher to the hospital. He did not want to be transported in an ambulance because, as he said, he did not want to catch lice. An officer on his deathbed and worried about catching lice!!"[7]

Private Amos E. Stearns, 25th Massachusetts Volunteer Infantry, as a prisoner of war at Florence, South Carolina, wrote a two-line poem on Jan. 16, 1865: "A Confederate prison is the place—Where hunting for lice is no disgrace."[8]

Lieutenant Sidney Carter, 14th South Carolina Volunteers, in a letter to his wife from near Berryville, Virginia, on Nov. 6, 1862: "I have to sleep with the crowd—and I am lousy already. I got lice off the men I brought from Richmond."[9]

Corporal Edmund D. Patterson, 9th Alabama Infantry, near Fredericksburg, Virginia, on Nov. 23, 1862: "Mr. Carroll says that he can't stay long, that these horrible lice will eat him up. He says that they are so thick that he is afraid to go to sleep, for fear that in an unguarded moment he might snore, and these vermin would think it was the dinner gong and eat him up."[10]

Private Galutia York, 114th New York Volunteer Infantry, in a letter to his family on Nov. 20, 1862, from Fort Monroe, Virginia: "the boys have had 2 or 3 battles we have come out victorious so far but I dont know how long we shall be so I tell you now it is quite a sight to see a thousand men drew up in a line of battle along the seashore all stripped of bearassed fighting lice boddy lice I mean them that is larger than a cornel of wheat with US under their tail and there is some that has got US marked on their backs there is one man in Co D that is so lowsy he is in danger of his life they got started with him the other night and got him pretty much out of the port hole before he got waked up I have not found but one on me yet but I expect to be covered with them before I get out of this for I cant help myself."[11]

Ticks (consisting of several genera and many species) are arthropods more closely related to spiders than insects such as body lice. Most species have three growth stages: larva, nymph, and adult. For the organism to progress from one stage to the next, it must have a blood meal in each case. The entire life cycle usually takes two years to complete. In addition to humans, ticks parasitize many other species of mammals, birds, and even some reptiles and amphibians. Being vectors of several human and animal diseases, they occasionally leave behind more than an itching bite. Lyme disease, Rocky Mountain spotted fever, tularemia, and relapsing fever are transmitted to people via ticks. Ticks are found in leaf litter and on plants. There is a common misconception that ticks jump or fall from plants onto a host. Actually, ticks are acquired only when they are physically touched.

Civil War soldiers in the field twenty-four hours a day had many opportunities to become unwanted parasitic hosts for dreaded ticks.

Captain Theodore A. Dodge, 101st New York Volunteers, near Chancellorsville, Virginia, on May 5, 1863: "Our great nuisance in sleeping on the ground here is the wood ticks. The Lieut. Col. has just found 4 stuck to him."[12]

Private Gottfried Rentschler, 6th Kentucky Volunteer Infantry [Union], in a letter to the *Louisville Anzeiger* from near Atlanta, Georgia, on July 6, 1864: "The weather has been unbearably hot since the 19th. If our whole army marched down through a narrow valley, the sweat that ran off us would form a river as large as the Mississippi, in which river the lice and wood ticks replace the fish and alligators; and all would drown miserably."[13]

Private Allen M. Geer, 20th Illinois Volunteers, near Corinth, Mississippi, on May 18, 1862: "Between wood ticks, gnats, & mosquitoes we had a lively time."[14]

Lieutenant Charles B. Haydon, 2nd Michigan Infantry, near Williamsburg, Virginia, on May 14, 1862: "We were considerably troubled by gnats & mosquitoes. Woodticks are however our greatest annoyance. It is impossible to keep clear of them."[15]

Colonel Francis C. Barlow, 61st New York Volunteers, in a letter to his brother from Yorktown, Virginia, on April 23, 1862: "We have exhausted wood ticks & have now bites from some mysterious animals or insects which we never see but know only from their bites."[16]

Private John Westervelt, 1st New York Volunteer Engineer Corps, at Folly Island, South Carolina, on April 18, 1863: "There is another nuisance here in the shape of the wood tick, an insect about the size and looking like the bed bug. They eat their way in the flesh without being felt for some time. It then begins to get sore and you look and find them half buried, and they are so tough you can hardly pull them out. It is necessary to examine yourself every day. If as is sometimes the case they eat all the way in it makes a bad sore."[17]

Continuing in the vein of misery-producing arthropods, chiggers or red bugs are the larval stage of a group of tick relatives known as harvest mites (*Trombicula* spp.). Harvest mites often occupy the same habitats as ticks but are microscopic in size. Like ticks, they parasitize a variety of mammals, birds, and reptiles. They do not suck blood but rather bury under the skin, inject the host with digestive enzymes, and ingest the resulting brew for nutrition. Unlike ticks, they are not usually vectors of disease in North America. Infections involving dozens of harvest mites on one victim are common and result in fiery itching sensations around the bites. Only the larval stage of harvest mites is parasitic; the adults feed on plant materials. Harvest mite populations fluctuate with environmental conditions such as temperature and humidity. During the Civil War, ideal circumstances for high populations and human exposure occurred toward the end of the Vicksburg campaign.

Lieutenant John Q. A. Campbell, 5th Iowa Infantry, near Vicksburg, Mississippi, in a letter to the [Iowa] *Ripley Bee* on May 22, 1863: "I send this note to let you know that I am still safe and well, but very dirty and nearly eaten up by chiggers. We have had no tents for four weeks, and have had to lie out in all sorts of weather. The chiggers are small (almost invisible) wood lice that are worse than fleas and gray backs combined."[18]

Private Theodore F. Upson, 100th Indiana Infantry Volunteers, near Vicksburg, Mississippi, on June 9, 1863: "We have fixed up pretty good quarters here with beds made of cane poles we get out of the cane brakes or swamps. There is one draw back to these cane beds; they are full of jiggers [chiggers], a little red insect which gets into the skin and makes one itch and scratch and some times they poison the men very badly."[19]

Captain Charles B. Haydon, 2nd Michigan Infantry, near Vicksburg, Mississippi, on July 1, 1863: "Nothing can exceed the beauty of the evening in this latitude. There is one great drawback. The infernal jiggers alias sand fleas eat the very flesh off one's bones. These are not the proper jiggers but a sort of red flea."[20]

Confederate sympathizer Myra Inman at Cleveland, Tennessee, on July 18, 1863: "Johnnie has a sore on his neck caused by a chigger bite. We are very uneasy about it, it is spreading on his face."[21]

DOLPHINS, PORPOISES, AND WHALES

Dolphins, porpoises, and whales are related groups of highly specialized, aquatic mammals found throughout the oceans of the world. Many species are found in the marine environments where most Civil War activities occurred, the Atlantic Ocean and Gulf of Mexico. In general, dolphins and porpoises are smaller than whales, and both are usually lumped under the term *porpoise* by laymen today as in the Civil War era. Scientists separate them based on anatomical differences—e.g., dolphins have cone-shaped teeth and short beaks, while porpoises have spade-shaped teeth and longer beaks. Whales are variable in size and include the blue whale (*Balaenoptera musculus*), thought to be the largest animal ever to have lived at more than one hundred feet long and weighing two hundred tons. Some whales, such as the sperm whale (*Physeter macrocephalus*), have teeth like dolphins and porpoises. Others have baleen plates that filter plankton and other small organisms from the water. Humpback (*Megaptera novaeangliae*), fin (*Balaenoptera physalus*), blue, and northern right whales (*Eubalaena* spp.) are in this group. Technically, orcas or killer whales (*Orcinus orca*) are considered a type of dolphin.

The American whaling industry was centered on the northeastern seaboard and employed thousands of seamen in the mid-nineteenth century. However, by the time of the Civil War the golden age of Yankee whaling had ended, in part as a result of the discovery of petroleum, which precluded the need for whale products such as lamp oil and lubricants. Also, populations of easily attainable whales in the Atlantic and Pacific Oceans had been overharvested, leaving abundant populations only in distant Arctic and Antarctic waters.

Like wars before it, the Civil War wreaked havoc on the whaling industry. Many northern whaling vessels were appropriated for some aspect of the war effort. Thirty-seven old whalers from the New Bedford area alone were sequestered to become part of the "Stone Fleet." They were filled with rocks and sunk in southern harbors, mainly Charleston, to impede southern shipping.[1] Most alarming for the industry was the fact that the Confederate navy and privateers considered their ships fair game in a wartime setting. Cruisers like the CSS *Shenandoah,* CSS *Alabama,* and the CSS *Florida* destroyed more than fifty Yankee whalers. That Union whalers were still impacting whales during the war is revealed in the cargos of captured ships. When the CSS *Alabama* captured the *Ocean Rover* in January 1862 off the Azores, it held 1,100 barrels of whale oil. The *Golconda* was laden with 1,800 barrels when torched by the CSS *Florida* off Bermuda in July 1864.[2]

References to whales in Civil War letters, diaries, and journals are uncommon. Whales tend to stay in deeper water away from coastlines where most shipping occurred. Many sources, however, mentioned "porpoises," which are often found near shore. Northern soldiers from the Midwest in particular wrote of them as novelties.

John Hay, assistant secretary to President Lincoln, near Fernandina, Florida, on Feb. 14, 1864: "We steamed out of the Harbor of Fernandina as the sun was rising. Struck about noon a school of whales—great shining black or dirty grey monsters with backs like hills & flukes like weavers' beams. I fired at one and missed him & felt unhappy."[3]

Private Theodore F. Upson, 100th Indiana Infantry Volunteers, off the coast of South Carolina in January 1865: "We are on a transport. The boys are amusing themselfs shooting at the porpoise—a big sea fish which are playing around our boat. The waves run pretty high and very few are able to hit them."[4]

John M. Follett, 33rd Illinois Volunteer Infantry, at Matagorda Bay, Texas, in a letter to his wife on Dec. 13, 1863: "I was out yesterday along the beach to see the Porpuss (I don't know how to spell it) and Dolphins play. The bay was full of them. They are a very large fish and floundered round in great style. I saw some as large as a horse and at least ten feet long."[5]

Corporal Rufus Kinsley, 8th Vermont Regiment, aboard the ship *Wallace* off the eastern seaboard on March 13, 1862: "Dead calm and all hands dead sick. Ship surrounded by schools of porpoise, highly tickled at sight of the Yankees, who looked, in their blueish hats and blueish green coats, as they were sprawling around the decks and hanging over the rails, quite like a cargo of overgrown bull-frogs."[6]

Colonel Thomas W. Higginson, 1st South Carolina [African American] Volunteers, off Cape Fear, North Carolina, on Nov. 23, 1862: "Now the soldier boys are in ecstacies at a school of porpoises & run lumbering from one side of the deck to the other to watch them."[7]

Private George A. Remley, 22nd Iowa Volunteers, in a letter to his brother from Brashear City, Louisiana, on Sept. 19, 1863: "Porpoises are frequently seen plunging about in the bayou. They seem to be huge monsters ten or twelve feet long and five or six feet wide and are perfectly harmless."[8]

Private John Westervelt, 1st New York Volunteer Engineer Corps, on board the steamer *Tappahannock* off the coast of Florida on Feb. 26, 1864: "This was the 26th and at noon we made land which proved to be Fernandina. Thousands of porpoises greeted us here and sported around the vessel for miles."[9]

John Hay, assistant secretary to President Lincoln, near Savannah, Georgia, on April 24, 1863: "We got away from Fort Pulaski at 5 oclock this morning. We skirted along within sight of land, among porpoises & Pelicans which were equally inaccessible till 3 P.M. when we entered the harbour of Fernandina."[10]

- Whale baleen plates were used to make hoops in women's fashionable skirts of the day.
- Sperm candle—"A candle made from whale spermaceti."[11]

FISH

More than nine hundred species of fish inhabit the various freshwater habitats of North America. Almost every stream, bayou, river, and lake harbors fish of some kind. The great diversity of species was generally unknown to the common person of the Civil War period, as the many varieties of obscure minnows, darters, chubs, and madtoms were seldom recognized. The situation is the same for the hundreds of types of fishes, large and small, associated with brackish and marine environments. Within the Civil War literature one thing is clear: fish were highly regarded as a source of food.

Soldiers sought fish at every opportunity across the broad landscape of the conflict. Also, the recreational value of fishing was important to many who suffered the boredom of army life. Methods of acquiring fish varied with the situation and creativity of the fishermen.

John S. Jackman, 9th Kentucky Infantry, in Jackson, Mississippi, on June 3, 1863: "The boys caught a great many fish out of the lake and [Pearl] river. One way of catching them was rather novel: Two men would go into the lake, when the water was not very deep, and hold a blanket spread out, down close to the water, then others would commence lashing the water about, making it muddy, and the fish would commence skipping above the surface of the lake, and fall on the blanket, thus being caught by hundreds."[1]

Major General Lafayette McLaws, CSA, in a letter to his wife from Fredericksburg, Virginia, on April 12, 1863: "The enemy are very quiet, and civil, our men not only fish in the river but seine it, catching hundreds of fish. The enemy fish with poles but have not ventured to seine as yet."[2]

Major James A. Connolly, 123rd Illinois Infantry, Murfreesboro, Tennessee, writing to his wife on June 9, 1863, of a raid on a Confederate cavalry camp: "They were encamped on the bank of a stream, and when we drove in their pickets some of their men were in the stream seining and had caught a fine lot of fish . . . the fishermen leaving their fish floundering on the bank."[3]

Assistant Surgeon Dr. Daniel M. Holt, 121st New York, at winter quarters on the Hazel River, Virginia, in a letter to his wife on Feb. 7, 1864: "The weather is balmy and springlike. Our boys go out, dig worms, and go to the river often returning with a string of nice fish.—I went one day, but as usual, got nothing. Fish do not like my bait any better than the men like my pills. Who can blame them?"[4]

Lieutenant John Q. A. Campbell, 5th Iowa Infantry, near Helena, Arkansas, on March 18, 1863: "The 'boys' are making traps to catch fish in the bayous. They have already caught some fine fish. The river is still rising."[5]

Corporal Robert A. Moore, 17th Mississippi Regiment, in Fredericksburg, Virginia, on May 13, 1863: "Have been seining in the Rhappahannock to-day. Have caught but few. The river is lined with fishermen. Shad, herring & perch are most abundant."[6]

Lieutenant Robert M. Addison, 23rd Wisconsin Infantry, on the Texas coast below Galveston on Feb. 6, 1864: "We had no drill this afternoon, so we went fishing. Co E and G had the use of the sieve [seine?]. Had good luck and caught quite a lot of fish."[7]

Private Galutia York, 114th New York Volunteer Infantry, in a letter to his brother on March 12, 1863, from a quarantine station below New Orleans: "Henry I wish that you could work it some way so as to send me some fishline and 2 or 3 hooks for there is a nice place hear to fish for there is lots of speckled trout."[8]

African American contraband William B. Gould as a sailor aboard the USS *Niagara* off the coast of Newfoundland on June 9, 1864: "made [saw] another sail, ran for it, A fisherman [fishing boat] at anchor boarded him and procurd A large lot of Fish, enough for all hands. We are now on the Grand Banks of Newfoundland."[9]

Lieutenant Sidney Carter, 14th South Carolina Volunteers, in a letter to his wife from near Fredericksburg, Virginia, on April 30, 1862: "I have eaten a fish caught here this morning—what they call a Stone Roller. It is only thirty feet from our tent to the water."[10]

Sergeant Henry C. Lyon, 34th New York Volunteer Infantry, near Elktown, Maryland, on March 10, 1862: "Boys trying the river for fish with good success. 'Abe' got us a fine mess of eels."[11]

Fish were important to civilians and other noncombatants as well.

George E. Stephens, African American servant to an officer in the 26th Pennsylvania Regiment, near Budd's Ferry, Maryland, on March 12, 1862: "Now that the rebel blockade of the Potomac has been raised the valuable fishing grounds will be accessible to the fishermen. The Potomac fisheries are worth thousands of dollars; and thousands of souls now in the throes of starvation will be furnished with food both savory and nutritious.... The fishing season commences about the 10th of March, at which time the gillers commence to take in the shad. About the last of March or the first of April seine-hauling commences. The shad season closes about the last of May; the seine-hauling about the 10th of May. Herring is the principle yield of these fisheries."[12]

British journalist William H. Russell at Washington, D.C., on March 28, 1861: "Dined with President [Lincoln] & his Cabinet ministers. Fish dinner."[13]

Kate Stone, Brokenburn Plantation near Milliken's Bend, Louisiana, on June 6, 1862: "We caught a pretty lot of fish out of the bayou just out in front of the house. Julia was the most successful fisherman."[14]

Mary Boykin Chesnut at Columbia, South Carolina, on May 24, 1862: "Before the war shut us in, Mr. Preston sent to the lakes for his salmon, to Mississippi for his venison, to England for his mutton and grouse."[15]

Harriett Goodwin Pierce, wife of Corporal Benjamin Franklin Pierce, 14th New Hampshire Volunteer Infantry, writing of her husband who was home on furlough in Bradford, New Hampshire, on April 22, 1865: "Frank and Len went fishing. Caught 39 fish."[16]

John Hay, assistant secretary to President Lincoln, near Hilton Head, South Carolina, on May 5, 1863: "The fish were mostly tangled in the net. Sam finds a porpoise in the net. Prawn. Angel fish. Trout. Bass. Flounder. Horse-shoe crab. Stingray. Cuttlefish. Silver eel. Garpike. Louse. Catfish. Look out for your fingers."[17]

The most common species of fish mentioned in surveyed Civil War letters, diaries, and journals was the American shad (*Alosa sapidissima*). Being anadromous, this species spends most of its life at sea and returns to freshwater rivers to spawn, usually only once in its life cycle. American shad are found along the Atlantic coast from Florida to the St. Lawrence River. During the Civil War they likely spawned in every river on the eastern seaboard. Their "runs" correspond with favorable water temperatures, usually between mid-February and early June.

One assessment blames American shad for the firing of a Confederate general. Referring to the Petersburg campaign, "The last named fish [shad] was partly responsible for bringing about the demise of at least one Confederate General during the siege. General George E. Pickett was attending a shad bake when his troops were overrun at Five Forks on April 1, 1865. Because he was not with his men while there was impending danger from the Federal Army, he brought the ire of General Lee who would eventually relieve him from command on April 8th."[18]

Remarks concerning American shad were common because they were considered excellent table fare, a commercial fishery for the species existed, they were often spectacularly abundant and accessible when spawning, and they occurred in areas having large troop concentrations, especially tidewater Virginia.

Private Richard H. Brooks, 51st Georgia Infantry, in a letter to his wife from Fredericksburg, Virginia, on April 27, 1863: "we are having our fun here a saining for fish an ketching fish with a hook an Lines we drag them out by the hundred with sains. They are mostly shad an herings, but the water is very cold yet."[19]

A. L. Peel, Adjutant, 19th Mississippi Regiment, near Fredericksburg, Virginia, on May 18, 1863: "Capt Dean took a detail of 10 men. Went saining in the river for the Regt. Caught 120 Shads."[20]

John Hay, assistant secretary to President Lincoln, near Fort Monroe, Virginia, on April 27, 1864: "We went down the river among the twilight 'shadders' and got some fish and dined off shad roe and shad."[21]

General Robert E. Lee in a letter to John Skaggs on March 29, 1863: "I am very much obliged to you for the pair of fine shad you sent me today. I acknowledge that these delicacies are very acceptable at a soldiers' dinner, but you know they are unnecessary & I fear will be much missed by the good citizens of Richmond. Our fare though simple is yet while thank God ample—I beg you will not deprive your self to administer to our comfort."[22]

Captain William J. Bolton, 51st Pennsylvania Volunteers, at Roanoke Island, North Carolina, on Feb. 17, 1862: "The fishing season is now in its prime and as there are plenty of fishing nets on the island the men hire them at a very small amount and the consequence is, we are having plenty fresh shad and herrings, and are living grand."[23]

Sergeant Taylor Peirce, 22nd Iowa Infantry, writing to his wife on Feb. 4, 1865, at Savannah, Georgia: "There has been fresh shad in the market but I have not been down to town this week although it is not 400 yards off."[24]

American shad were a commodity, and prices were subject to their availability.

John B. Jones, clerk to the Confederate secretary of war, in Richmond, Virginia, on April 27, 1864: "Food is still advancing in price; and unless relief comes from some quarter soon, this city will be in a deplorable condition. A good many fish, however, are coming in, and shad have fallen in price to $12 per pair."[25] **Less than a year later, with the fall of Richmond imminent, Jones wrote on March 29, 1865:** "Shad are selling at $50 per pair."[26]

Private George M. Englis, 89th New York Volunteers, in a letter to his family from Roanoke Island, North Carolina, on March 12, 1862: "We have lots of fresh shad here. I can buy a big one for 15cts."[27]

Colonel Charles F. Johnson, [Union] Invalid Corps, in a letter to his wife from Belle Plains, Virginia, on May 22, 1864: "the most that troubles me is the eating, it is true that we get 'fresh shad' caught at the very foot of our camp but 'nary conscience' they only charge just one dollar per foot for them, that is, a shad 12 inches long is $1.00, larger ones in proportion."[28]

Catfish were another group of fish frequently mentioned. Although widespread in range, they are most abundant in the South. Three species—

flathead catfish (*Pylodictis olivaris*), channel catfish (*Ictalurus punctatus*), and blue catfish (*I. furcatus*)—grow very large, and individuals exceeding one hundred pounds were common before overexploitation. Then as now, catfish filets were considered a delicacy by many, but not all, people.

Private Isaac Jackson, 17th Ohio Battery, during the siege of Vicksburg, Mississippi, on June 28, 1863: "There are plenty of fish at the river now, but very few get this far out. I seen some of the biggest fish today I ever saw. One man had a cat fish on his shoulder, it was about as much as he could carry."[29]

Lieutenant John Q. A. Campbell, 5th Iowa Infantry, near Helena, Arkansas, on Sept. 22, 1863: "Fishing is getting to be quite a business with us. Many large catfish have been drawn from the 'Father of Waters' in a few days past."[30]

Lieutenant Robert M. Addison, 23rd Wisconsin Infantry, at Mobile Bay, Alabama, on Aug. 30, 1864: "Finished the roll then went fishing, had fine luck caught some splendid cat fish. Came in at noon, had our excellent dinner of fish."[31]

Captain Charles B. Haydon, 2nd Michigan Infantry, at Memphis, Tennessee, on June 15, 1863: "The soldiers hauled in a great many big bellied, lubberly catfish weighing from 20 to 40 lbs. each and for further amusement rolled the female contrabands about all ends uppermost."[32]

Private Alexander Downing, 11th Iowa Infantry, at Vicksburg, Mississippi, on Jan. 14, 1864: "Some of our guard early this morning stole a big fish from a fisherman who was taking a load to market. It weighed forty pounds and was divided among the boys. I took a piece to my tent and cooked it, but I might as well have eaten a piece of crow, for it was tasteless and tough. It proved to be a channel cat."[33]

The alligator gar (*Atractosteus adamantinus*) is one of the largest species of freshwater fish in North America. Primitive in appearance with a formidable snout full of teeth, some individuals weigh up to three hundred

pounds. The reputation of this species as dangerous to humans is not jus-
tified, although a very few injuries caused by gars have been documented.
Three other smaller species of gars occur in the United States.

**Private Isaac Jackson, 83rd Ohio Volunteer Infantry, near New Orleans,
Louisiana, on Aug. 16, 1864:** "There was quite an accident happened to
one of Co. F today. He was down to the river washing his shirt. He was
standing at the water's edge washing, when a 'Gar' came up and caught
hold of his hand. It nearly cut three of his fingers off. It nearly jerked his
arm off, he said. The Alligator Gars are a savage looking fish. They have a
very large mouth with a long bill running out in front. They look like they
could take a man's leg off at one snap. I will send you the scales of one I
found on the river bank. It was only 18 or 20 inches long."[34]

**Private Theodore F. Upson, 100th Indiana Infantry Volunteers, near
Vicksburg, Mississippi, on June 9, 1863:** "It is very warm, and the boys
will go swiming in the River which is not good for them. Strict orders
have been given to stop it, yet they will go. Yesterday a lot were in and one
started to swim across the River. Just as he got out in the deep water he
threw up his ha[n]ds, gave a yell and went under and did not come up. The
Darkys say a gar got him. Just what a gar is we do not know but the boys
don't go swimming any more."[35]

**Lieutenant John G. Earnest, 79th Tennessee Infantry, near Vicksburg,
Mississippi, on May 5, 1863:** "Went out to Chickasaw Bayou on picket
this evening. As soon as we got the men posted, I took a fish—had some
glorious nibbles and could see alligator gars jumping up all around me
but only succeeded in catching a small perch. Some of the gars jumping
around in the water were five or six feet long and would probably weigh 20
pounds."[36]

**Fleeing Confederate refugee Sarah Wadley near Bayou Lafourche, Lou-
isiana, on Sept. 23, 1863:** "This is our Second Camp. Here is "Lazy Pete"
with two gar fish in his hands, which he has just caught in a mud hold near
by, they look like alligators with the large mouths and feet like fins, and
long pointed tail. Pete says 'they taste like any other fish' 'boil 'em and fry
'em and bake 'em' but I don't think I should like to taste one."[37]

꙳

Like gars, sharks are top predators in their ecosystems and are often persecuted. Many of the dozens of species worldwide have been greatly reduced in numbers since the Civil War. Soldiers and sailors ate sharks occasionally and respected them no less than mariners today.

Private John Westervelt, 1st New York Volunteer Engineer Corps, at Folly Island, South Carolina, on Aug. 13, 1863: "I had a new dish to day. It consisted of shark steak. One of our company while fishing to day caught [one] about 5 ft long. . . . The meat looks beautiful but is coarse."[38]

Surgeon's Steward C. Marion Dodson aboard the USS *Pocahontas* off Sabine Pass, Louisiana, on June 16, 1864: "One of the men caught a young shark. Twas a good flavor as we all enjoyed a piece of him."[39]

Corporal Rufus Kinsley, 8th Vermont Regiment, aboard the ship *Wallace* off the eastern seaboard on March 14, 1862: "Two men in Co. I very sick with lung fever: not likely to live long. A huge shark alongside, to save them from being buried 'at the bottom of the deep blue sea;' whereat they are doubtless very grateful, if they are aware of the shark's disinterested benevolence." **Twelve days later he wrote:** "Caught a shark 8 feet long, with a hook. He fought like a tiger, and came near swallowing the mate; but the carpenter's ax made a bad hole in his head, after which he kept very quiet while surgeon Gale dissected his body. Saw two whales."[40] **Later, as a lieutenant in the Second Corps d'Afrique, he wrote his father from Ship Island, Mississippi, on May 29, 1864:** "I was very near furnishing a shark with breakfast one morning last week. We were but a short distance from the end of the wharf, in eighteen feet of water, when I, being farther out than the others, excited the appetite of the ravenous man-eater, who at once evinced his anxiety for an intimate acquaintance. I made for the wharf, and was so fortunate as to reach it in time to give a fisherman on the wharf opportunity to throw his gig. He failed to strike the creature, but frightened him away. Before we left the wharf we saw a desperate fight between the shark and a school of porpoise. The shark was vanquished after a fight of about two hours duration. Many sharks are taken here at this season of the year."[41]

Soldiers often mentioned species new and interesting to them, and some even discovered that they were considered "fish" themselves.

Surgeon's Steward C. Marion Dodson aboard the USS *Pocahontas* off the eastern seaboard on April 27, 1864: "First time I ever saw flying fish. Seemed quite curious to us all. Also saw Nautalus. They would spread out their web sails and really look like a miniature craft under way."[42]

John King, 40th Georgia Infantry, on his first day at Camp Chase Prison in Columbus, Ohio: "The light of day had come, and with its coming the old prisoners, who were awakened from their slumbers of the night, seeing us walking over the prison grounds began to exclaim 'Fresh Fish.' The refrain came back from many others who, aroused by the cry, were coming from their bunks in the prison pews, anxious to see the 'Fresh Fish.' Very naturally we looked about to see some monger who was bringing in 'fresh fish' for our breakfast, and began to feel our appetites sharpened in anticipation of a feast after our long fasting. We were soon undeceived, as we discovered that we were the 'Fresh Fish.'"[43]

FLIES AND MOSQUITOES

Of all the species in the animal kingdom, none had more impact on the Civil War than those in the taxonomic insect order *Diptera*. Characterized by a single pair of wings, this group includes mosquitoes, houseflies, black flies, sand flies, horse flies, gnats, midges, and many more. About 120,000 species have been described, with estimates of more than one million species worldwide. The basic life cycle involves egg, larva (maggots in some species), pupa, and adult.

While some kinds of flies are important plant pollinators, this group of insects is renowned for the misery and suffering it caused to Civil War participants. The irritating bites of hordes of blood-sucking parasites such as mosquitoes, sand flies, and black flies were overshadowed by disease and death caused by these species in their roll as vectors of pathogens. Common houseflies (*Musca domestica*) can transmit poliomyelitis, typhoid fever, tuberculosis, anthrax, leprosy, cholera, diarrhea, dysentery, and conjunctivitis.[1] Although the relationship was not discovered until after the Civil War, the plagues of malaria and yellow fever were borne on the wings of mosquitoes. Citizens, soldiers, horses, and mules all struggled to survive this second war with the armies of insect order *Diptera*.

Colonel Thomas W. Higginson, 1st South Carolina [African American] Volunteers, at Port Royal Island, South Carolina, on April 17, 1863: "not even dreams had prepared me for sandflies. Almost too small to be seen they bite worse than musquitos & leave more lasting soreness. They are almost incompatible with Dress Parade; fancy me standing motionless, with my face a mere nebula of these little wretches, torrents of tears rolling down my expressive countenance, from mere muscular irritation. They are however a valuable addition to discipline as they abound in the guard house and render that institution an object of unusual abhorrence. Thus do the weak things of the earth confound the mighty."[2]

Reuben A. Pierson, 9th Louisiana Infantry, in a letter to his sister from Camp Moore, Louisiana, on June 20, 1861: "There are more flies in and around Camp Moore than there are in all Bienville Parish. . . . It would turn the stomach of any other being except a soldier to go into one of the eating houses kept on this encampment. The flies are so thick until you have to be careful in carrying a mouthful from your plate to your mouth lest a fly should alight upon it before it is received."[3]

Private George A. Remley, 22nd Iowa Volunteers, in a letter to his mother from near Vicksburg, Mississippi, on June 9, 1863: "I am sitting by Lycurgus [his brother] keeping the flies off him with one hand and writing with the other, just now, however, Lycurgus offered to do that for himself until I get through writing."[4]

Private Wilbur Fisk, 2nd Vermont Volunteers, near Warrenton, Virginia, in a letter to his hometown newspaper on Aug. 13, 1863: "Flies, too, are becoming miraculously abundant, and as annoying as they are abundant. We have the most prolific species of this insect here that I ever saw. I believe they increase fourfold every day, and have done so in regular geometrical progression ever since dog days commenced. They swarm everywhere and torment a fellow from daylight till dark. They hover in his face, fly into his mouth and nose, and with their little tickling feet irritate the flesh wherever the saucy imps can find it. There seems to be a peculiar species of the fly kind that infests our camp. They bite almost as quick as a bee can sting, and their bite is almost as painful. The rebels could never boast of being more troublesome than these same little flies. It would be untrue to say that we are idle; we cannot be in such a nest of flies."[5]

Sergeant John Q. A. Campbell, 5th Iowa Infantry, near Jacinto, Mississippi, on Aug. 11, 1862: "Some of the boys on grand guard yesterday had their blankets and clothing fly-blowed! Jiggers and flies are the plagues of Tishomingo County."[6]

Lieutenant Henry C. Lyon, 34th New York Volunteer Infantry, near Malvern Hill, Virginia, on Aug. 7, 1862: "After guard mounting nothing on program until dress parade then supper then went for mail. all this time, however we constantly fight flies."[7]

Private George A. Remley, 22nd Iowa Volunteers, in a letter to his brother from near Vicksburg, Mississippi, on July 4, 1863: "The musketoes are not very bad but the flies are so numerous that they will eat up a pint of sugar in a half a day if it is left uncovered. This is a fact!"[8]

John M. Follett, 33rd Illinois Volunteer Infantry, at Brasheur City, Louisiana, in a letter to his parents on June 19, 1864: "It is anything but pleasant here at this time of year. it is hot all day and the flies, fleas, spiders, crickets, slugs, beetles, giggers, lice and all kinds of vermin bother so one has no peace or quiet and at night the frogs, owls, aligators, musquitoes, and all creeping things keep up such an ever lasting hissing and noise it is almost impossible to sleep. The musquitoes present their bills for liquidation, the fleas play backgammon on my anatomy all night and I wake up in the morning more tired than when I laid down."[9]

The influence of this group of insects on the war must also be considered in light of the extensive damage caused to food supplies.

Colonel Charles F. Johnson, 81st Pennsylvania Volunteer Infantry, in a letter to his wife from near Fredericksburg, Virginia, on Nov. 17, 1862: "I have not had meat for two days and the hard crackers are those which were originaly shiped to the Peninsular and back and are actualy full of Bugs and white worms simlar to maggots, except that they have black heads— it is horrable to contemplate, but hunger and costume can make one use to almost any thing and we actualy eat them with a full knowledge that they are there; but it can only be done by biting into them and chewing away without looking at them."[10]

Confederate surgeons apparently used fly maggots to clean wounds at times. A Union surgeon reported: "I saw several gangrenous wounds filled with maggots. I have frequently seen neglected wounds amongst the Confederate soldiers similarly affected, and, as far as my experience extends, these worms destroy only the dead tissues and do not injure specially the well parts. I have even heard surgeons affirm that a gangrenous wound which had been thoroughly cleansed by maggots healed more rapidly than if it had been left to itself."[11]

Captain John Dooley, 1st Virginia Infantry, as an injured prisoner of war in Fort McHenry, Maryland, on July 17, 1863: "Today Sheppard who is most kind in his attentions to my wounds extracts therefrom 4 maggots and cleanses the wounds thoroughly. They are doing very well now; I mean my wounds."[12]

Private Melville Follett, 42nd Illinois Infantry, in Georgia on Sept. 24, 1863: "Had an awful night of it last night. We are lying on the naked ground and I became so worn out that I thought I could not live until morning. My wound is very troublesome and gives me more pain than I can tell. Smith of my company is on my left and he discovered that he was completely covered with maggots. Poor fellow how he suffered."[13]

Mosquitoes were most abundant in their natural haunts of coastal marshes and inland swamps of the Mississippi Valley. Soldiers reported their aggravation frequently.

William L. Nugent, 28th Mississippi Cavalry, in New Orleans, Louisiana, in a letter to his wife on April 10, 1862: "The mosquitoes are becoming somewhat annoying already, and will trouble us exceedingly unless we can get netting enough to make us bars. Please ask Ma to get the netting at Penrice's and make a bar for me and Clarence immediately."[14]

Private Isaac Jackson, 83rd Ohio Volunteer Infantry, at Baton Rouge, Louisiana, on June 28, 1864: "The 'skeeters' here are—well, there is no use talking. You never can imagine the like—I never seen the like before."[15]

Lieutenant John Q. A. Campbell, 5th Iowa Infantry, near Vicksburg, Mississippi, on June 29, 1863: "I was cheated out of my sleep by the mosquitoes last night, getting only a fifteen minute snooze during the whole night."[16]

Sergeant George W. Bolton, 12th Louisiana Volunteers, in a letter to his family from near Memphis, Tennessee, in May 1862: "They [mosquitoes] are as bad as the buffalo nats on your mules in the spring. They are not so bad in the tents as they are in the woods because the camp fires and the smoke keeps them off but just get into the woods early in the morning or late in the evening, we can hardly stand them."[17]

Private Nelson Stauffer, 63rd Illinois Infantry, near Cairo, Illinois, on July 1, 1862: "On guard. Mosquitoes very bad bit us through a wool blanket."[18]

Major Charles F. Johnson, [Union] Invalid Corps, in a letter to his wife from St. Louis, Missouri, on Aug. 14, 1863: "As it is now striking 11 o'clock and the mosquetoes have discovered my light I will close—we are compelled to sleep under mosqueto nets here for they are thick enough to darken the street lamps."[19]

Colonel Lafayette McLaws, 10th Georgia Infantry, in a letter to his wife from Fort Magruder, Virginia, on July 30, 1861: "The weather is excessively warm and the mosquitoes very active all the day long, but I have on tar & when in bed am independent."[20]

Sergeant William D. Dixon, [Savannah] Republican Blues, at St. Catherine's Island, Georgia, on Sept. 12, 1861: "This has been a very warm day and tonight there is not a breath of air and the misquitoes are dreadful bad. We had to make a pine straw fire in front of the tents to smoke them out before we could stand in them."[21]

Private Allen M. Geer, 20th Illinois Volunteers, near Lake Providence, Louisiana, on March 20, 1863: "The mammoth Southern Mosquitoes are getting too thick to be agreeable here."[22]

Private George A. Remley, 22nd Iowa Volunteers, in a letter to his brother from Bayou Boeuf, Louisiana, on Sept. 8, 1863: "Musketoes, gallinippers, crab fish, gars and alligators are some of the luxuries of this country."[23]

Mosquitoes were prime subjects for literary exaggeration, and soldiers were often up to the task.

Corporal George M. Englis, 89th New York Volunteers, in a letter to his mother from Folley Island, South Carolina, on Aug. 7, 1863: "Nights we have to listen to the *sweet melodies* of the Mosquito and nothing but an Iron Clad is proof against their persevering bills."[24]

John S. Jackman, 9th Kentucky Infantry, at Vicksburg, Mississippi, on July 21, 1862: "At night our company deployed on the [Miss.] riverbank.... While in camp, the mosquitoes never bothered us on account of the camp-fires, but when we would get out on duty like this, where we were not even allowed to smoke a pipe, the mosquitoes would give us 'fits.' I had often heard of their being so large in Texas that they carried a brick-bat under their wings with which to whet their bills, but I never believed this story, until I came to Vicksburg."[25]

Lieutenant John G. Earnest, 79th Tennessee Infantry, near Vicksburg, Mississippi, on May 5, 1863: "A huge old mosquito with claws like a ground hog and a bill half as long as a sergeant's sword was seated on my shoulder trying to run his bill through my neck and pin me to the ground—fortunately it had lodged against my backbone and before he could make another trial the sentinel came to my relief at a 'charge bayonets'—when 'old skeeter' flew off saying he 'would be happy to repeat the call.' I sincerely hoped he would not. I now set about tucking the cover under me all around and finally went to sleep again. About daylight I was awakened by a tremendous roar—when I found the mosquitoes had pulled me to the edge of the bayou, and an old alligator jubilant at the prospect of getting me for his breakfast had given a tremendous laugh which awoke me, and I preferring not to be his breakfast shifted from there. I vowed never to allow myself to sleep on that bayou's bank again."[26]

FROGS

About fifty species of frogs and toads are found in the eastern United States. They live in a diverse range of habitats, from below ground to treetops. Most are closely tied to wetlands during part or all of their life cycle, especially during egg and larval (tadpole) stages. Frogs occupy important roles in the middle of food webs. They eat insects and other invertebrates and are eaten by fish, snakes, egrets, herons, owls, raccoons, mink, and humans, among others.

Frogs were likely more abundant during the Civil War era than today because most wetland habitats were still intact. In the period since, large-scale habitat losses have resulted from activities such as agriculture in the lower Mississippi River Valley and pine monoculture farther east. Smaller yet detrimental losses were the result of urban sprawl that often fragmented and isolated important habitats. Scientists consider frogs and other amphibians as biological indicators of overall ecosystem health because of their permeable skin that is sensitive to pollutants, their intermediate position in the food web, and the fact that many species spend part of their lives on land and part in water. This leads to the premise that collapsing frog populations should raise concerns for overall ecosystem (including human) health.

The idea of frogs as "coal mine canaries" was foreign during the Civil War. Most soldiers mentioned frogs only incidentally in their letters, diaries, and journals. Some considered the sound of calling frogs a pleasant experience; others did not. Perhaps the contrast was a result of an individual's state of affairs at the time. The first two anecdotes reveal the differences in attitude as a northern and a southern soldier write about frogs on the same Georgia night while only ninety miles apart.

Sergeant Taylor Peirce, 22nd Iowa Infantry, writing to his wife on Feb. 4, 1865, at Savannah, Georgia: "The weather is mild and pleasant. The frogs are Singing and all nature shows the near approach of Spring."[1]

John S. Jackman, 9th Kentucky Infantry, near Waynesboro, Georgia, on Feb. 4, 1865: "Our camp is on the border of a little lake, out of which we use water, and which abounds with frogs. The frogs keep up such a croaking as to prevent us from sleeping at night."[2]

Private Isaac Jackson, 83rd Ohio Volunteer Infantry, at Baton Rouge, Louisiana, on June 28, 1864: "I had a nice little scare last night on picket. After I was relieved at 11o'clock, I went to my bed and got in. . . . Was lying on my side when I felt something getting on my back. I turned over slowly to see what it was. I seen it and thought it was a snake, sure, and you ought to have seen me jump. It would have made you laugh, I am sure. I nearly jumped from under my bar. I took a second look and found it a harmless toad which went hopping off when I raised my bar. But that scared me out of 2 hours sleep at least."[3]

Private Samuel P. McKee, 22nd Georgia Volunteers, in a letter to his brother from near Portsmouth, Virginia, on March 10, 1862: "The weather is nice. The frogs is hollowing and spring birds is singing and it looks like corn planting time."[4]

Rev. Francis Springer, chaplain, 10th Illinois Cavalry, at Fort Smith, Arkansas, on Sept. 4, 1863: "To reach this place has cost our little Army of the Frontier a toilsome march of 350 to 400 miles. For many days the atmosphere was a flame of heat & dust; our best drinking water was that of ponds & puddles, the native residence of sturdy & senator-like frogs surrounded by their admiring crowds of musketo wigglers, & tadpoles."[5]

James T. Ayers, 129th Illinois Volunteers, in Hardeeville, South Carolina, on Jan. 17, 1865: "This is A singular Country timbers all Evergreen makes it Look just like Spring or summer and then thare are frogs plenty here and they are singing Just as merily now as they do in the Spring season in Illinois or Indiana and seem Just as merry as Tho thare was no war in our Land."[6]

Private Galutia York, 114th New York Volunteer Infantry, in a letter to his parents on Feb. 25, 1863, from a quarantine station below New Orleans: "I understand that we start for the reg this week some time if nothing happens and I hope and pray that we shall go for I am tired of staying hear for it is a low wet hole inhabited by crockadiles frogs mosquitoes and rats."[7]

Private Amos E. Stearns, 25th Massachusetts Volunteer Infantry, at Newport News, Virginia, on March 12, 1864: "It was a very fine day and the frogs are peeping tonight. No letters today."[8]

Private Nelson Stauffer, 63rd Illinois Infantry, near Pee Dee River, South Carolina: "The country looks miserably poor, almost too poor to furnish grub for the frogs and tadpoles."[9]

Jedediah Hotchkiss, topographical engineer of the Army of Northern Virginia, near Fredericksburg, Virginia, on March 9, 1863: "It was quite pleasant; frogs and birds singing,—turtles crawling out in the sun."[10]

Private William R. Stilwell, 53rd Georgia Volunteers, in a letter to his wife from near Richmond, Virginia, on July 22, 1862: "when the moon is up in the heavens and the gentle wind from the mountains sends forth its still rustlings among the aspen tree under which I stand while the thousands of rattles of the drums are all still and the frogs that sing in the swamp all around are sending forth their beautiful melody while I go from post to post with a little black box in my hand containing the moments of my soul with a little lock of golden braid."[11]

- Frog—a looped leather scabbard for a sword or bayonet.
- "Frog in the Well"—a Civil War fife and drum song.

HONEYBEES

For thousands of years humans have gathered honey from the hives of wild honeybees (*Apis mellifera*). People ate honey, concocted the alcoholic beverage mead, and made candles from beeswax. Honey and wax were also used for medicinal purposes. None of this, however, happened in North America until Europeans arrived, because honeybees are not native to the Western Hemisphere. Records indicate that honeybees were shipped from England to the Colony of Virginia in 1622. Other shipments were made to Massachusetts around 1630. Swarms of these early colonies soon escaped and became the "wild" honeybees of North America.[1]

For centuries, colonies of honeybees have been kept in wooden boxes, straw skeps, and pottery containers. In America pioneers kept bees in "bee gums," sections of hollow logs that were used as hives. The black gum tree was a favorite because of its tendency to form hollows. Collecting honey from any of these hive types usually resulted in the death of the bees and loss of the colony. A few years before the Civil War, a Pennsylvania minister patented a hive with movable frames that is still used today. It allowed collecting the honey without loss of the colony. C. P. Dadant, a French immigrant in western Illinois, began selling these modern beehives and frames to his neighbors in 1863.[2] Still, during the Civil War period most honey was gathered by cutting bee trees and taking what honey was available, and by killing colonies and taking the honey from kept hives.

Beeswax was an article of commerce soon after it became available in the colonies. Widely used for candles, the wax was melted, poured into molds, and transported to market. Confederate surgeon Francis Porcher in his Civil War treatise lists the following recipe for wartime candles: "take one pound of beeswax, and three-fourths of a pound of rosin, melt them together, then take about four threads of slack-twisted cotton for a wick, and draw it about three times through the melted wax and rosin, and wind it in a ball; pull the end up, and you have a good candle."[3] The demand for candles in the South during the war is recorded in the following passage: "Mr. B. Metcalf, of Montgomery, relates that he attended an [blockade runner's] auction sale, at Mobile, on one occasion, and, arriving late, found the cargo all sold except cod-liver oil and bees wax, which he succeeded in purchasing. His two barrels of cod-liver oil and 600 pounds of bees-wax were immediately reshipped to Montgomery on the Alabama River. Filling every shape and size bottle to be found, and placing a judicious advertisement in the papers, he was enabled to sell the oil, but what to do with the bees-wax was a puzzler. Discovering a set of candle moulds and using cotton yarn as a wick, he ran the entire mass into candles and succeeded in selling the whole stock at ten cents apiece."[4] An Alabama woman who coped with the trials of the blockade wrote after the war, "When beeswax was plentiful it was mixed with tallow for moulding candles. Long rows of candles so moulded would be hung on the lower limbs of wide-spreading oaks, where, sheltered by the dense foliage from the direct rays of the sun, they would remain suspended day and night until they were bleached as white as the sperm candles we had been wont to buy, and almost as transparent as wax candles."[5] As for Civil War medicines, one source reports on the commonly used drug blue mass as follows: "Civil War surgeons used this poisonous compound of mercury, chalk, licorice, and honey almost indiscriminately."[6]

The products of honeybee colonies, wild or in kept hives, were among the prizes most sought after by foraging soldiers. Sweeteners of any kind were at times scarce during the war, and honey was always considered a treat, often to the detriment of beekeepers.

Lieutenant John P. Sheffey, 8th Virginia Cavalry, in a letter to his future wife from Fayette County, Virginia, on Nov. 3, 1861: "Then a determined squad went forth, true votaries of Mercury, the god of thieves, and pressed two Union bee gums. The gums came in laden with stores of honey & the honey comb rich as ever grew in Hybla or on Hymetus. . . . A few poor luckless wights [*sic*] were stung. What mattered it? One Border Ranger, a *rara avis* on all occasions, eat bees and all. One stung him on the lip & to have revenge he bit the bee & was stung in the mouth & then, enraged, he 'bolted' him head, honey, sting and all."[7]

Private Theodore F. Upson, 100th Indiana Infantry Volunteers, near Atlanta, Georgia, in November 1864: "We came along by a big house close by the road that had a porch or recess in the upper story. This was full of bee hives and some of the boys were tumbling them down to the ground. The bees were pretty thick but we got plenty of honey."[8]

Lieutenant William R. Montgomery, 1st Georgia Sharpshooters, near Hagerstown, Maryland, on July 9, 1863: "Went last night with four men to guard an old man's Bee Gums. Gave us a good breakfast."[9]

John S. Jackman, 9th Kentucky Infantry, at Vicksburg, Mississippi, on July 25, 1862: "About sundown, a neighboring bee-gum was brought in, and we had a 'sweet time.'"[10]

Private Henry R. Berkeley, Amherst Battery in the Army of Northern Virginia, near Sperryville, Virginia, on July 25, 1863: "I got a nice breakfast at a little house on the side of the mountain near Sperryville, consisting of an abundance of nice hot biscuits, nice butter, rich milk and a plenty of nice strained honey. Paid a dollar a piece for it. I never enjoyed a breakfast more."[11]

Confederate sympathizer Myra Inman at Cleveland, Tennessee, on Sept. 7, 1864: "I ate some honey; it made me sick."[12]

Assistant Surgeon William Child, 5th New Hampshire Volunteers, in a letter to his wife from near Petersburg, Virginia, on Oct. 24, 1864: "I wish you would give me a long account of how you and our children get on. I want to hear how many chickens you have—and how the bees have done during the summer. I have not heard a word of them."[13]

Sergeant Alexander Downing, 11th Iowa Infantry, near Fayetteville, North Carolina, on March 14, 1865: "I went out early this morning with the foraging party. . . . We came to a rich plantation about four miles out. . . . After the boys had caught and loaded all the chickens and upset fully a hundred beehives, they called out, 'The rebels are coming!' . . . It was every fellow for himself."[14]

Gideon Lincecum, seventy-one-year-old naturalist, in a letter to a friend from Washington County, Texas, on May 23, 1864: "The birds are eating up my bees, and for want of the [percussion] caps on my part, they will be able to continue their destructive work."[15]

LIZARDS

At least seventeen native species of lizards and their kin are found in the eastern United States. Since the Civil War, almost as many exotic species have been introduced, primarily in south Florida. A great many more are found in the West, especially in areas with warm dry climates. Lizards were mentioned only incidentally in Civil War writings and were usually considered vermin, in the same category as snakes.

Confederate nurse Ada Bacot near Charlottesville, Virginia, on July 18, 1862: "after tea I was sitting here by the fire with no other light in the room, when I saw a little black somthing run out from under the bed, & as quickly run back again, I thought it must be a lizard so I ran out & called Savary & one of the other servants & made them move the bed & look every where they found nothing so I thought it must have escaped through a hole in the window shuter, how I am to sleep I cant tell, I have even a greater horror of a lizard than a snake."[1]

Sergeant George A. Remley, 22nd Iowa Volunteers, in a letter to his father from Matagorda Island, Texas, on March 24, 1864: "Snakes, lizzards and many such creeping things are numerous. There is one the boys call a horned toad, but it does not resemble a toad in the least. It does not hop but runs along the ground. It is five or six inches long, the body three or four inches across has a tail and legs resembling those of the lizard. Its peculiarity is several sharp horns, like thorns, on the top of the head and the rough jagged appearance generally. It is tame & perfectly harmless."[2]

Captain Jacob Ritner, 25th Iowa Infantry, in a letter to his wife from near Acworth, Georgia, on June 7, 1864: "But there are several things that make a night's sleep down here, not just so pleasant as it might be. For instance, we always have to put our pants inside of our stockings at night to keep the lizards, spiders, bugs and ants, &c from crawling up our legs!"[3]

Private William H. Bradbury, 129th Illinois Infantry, in a letter to his young daughter from near Atlanta, Georgia, on July 11, 1864: "We see lizards every day. They are about six inches long and live about old logs where they feed upon insects. They don' harm anybody. I have not seen a snake yet. I saw some lizard's eggs the other day. These eggs make young lizards."[4]

MOLLUSKS

Mollusks are a very diverse group of invertebrates that includes snails, clams, oysters (addressed separately in this book), squids, and octopus. Most live in the oceans of the world, but many are found in fresh water and on land. For centuries, humans have exploited certain species of mollusks for food. Others are used as a source for pearls, mother of pearl, and dye. Seashells are the protective structures excreted and occupied by some mollusks. Collecting seashells was just as popular during the Civil War as today.

John M. Follett, 33rd Illinois Volunteer Infantry, at Matagorda Bay, Texas, in a letter to his wife on Dec. 13, 1863: "I picked up some shells yesterday and when I came in I sorted out the best ones and put them in a box and shall send them. . . . In the box I send a sample of sea weed, a sea crab claw (the blue long one) a lobster claw, some specimens of coral a specimen of oyster formation, very small, and several other shells."[1]

Captain William J. Bolton, 51st Pennsylvania Volunteers, at Roanoke Island, North Carolina, on Feb. 12, 1862: "A most beautiful day. Wrote several letters and sent home a box of shells I had picked up along the beach."[2]

Lieutenant Robert M. Addison, 23rd Wisconsin Infantry, on the Texas coast below Galveston on Jan. 30, 1864: "No battalion drill this afternoon many of the boys went fishing and I went to the sea shore and gathered some shells to send home."[3]

Conchs are a group of large saltwater snails valued around the world as epicurean delights. Omnivorous Civil War soldiers on the Atlantic coast ate them opportunistically.

Private John Westervelt, 1st New York Volunteer Engineer Corps, at Folly Island, South Carolina, on April 23, 1863: "Do you recollect the conk shells up to grandpops that they used to blow for dinner well there are plenty of them here alive and we consider them a great dish."[4]

Sergeant William D. Dixon, [Savannah] Republican Blues, at St. Catherine's Island, Georgia, on Aug. 26, 1861: "We walked about 3 miles, gathering some Conks and brought them to camp. It is very good walking on the beach on the south side for it is fully exposed to the open sea."[5]

⤷

Idle soldiers, including prisoners, often spent their time making an assortment of jewelry and curios from whatever natural materials were at hand. Shells from freshwater clams (mussels) were a popular raw material.

Captain William H. A. Speer, 28th Regiment of North Carolina Troops, as a prisoner of war at Johnson's Island in Lake Erie on June 22, 1862: "Various amusements are followed by the prisoners. The most are employed in ring making in which they use guts [?] & such for the ring & put in it sets of gold & silver but mostly of shell of various colors, some of which are exceeding nice. The shells are obtained out of Lake Erie. We get them when we go into the lake bathing by diving after them. Many thousands of the rings, breastpins, shirt buttons, bracelets & watch fobs are made."[6]

Private Harvey Reid, 22nd Wisconsin Volunteer Infantry, in a letter from Murfreesboro, Tennessee, on Aug. 3, 1863: "The camp, since we have been here has been a regular manufactory of trinkets, and every one who had the genius or taste for such employment had some shell ornament to send home. At Annapolis, the boys' material for displaying that ingenuity was laurel root and cannel coal—at St. Louis, everything was made of bone—here they use clam shells from the river. The fact of the shells being found in Stone River is supposed to give the trinkets great additional value."[7]

OPOSSUM

The opossum (*Didelphis virginiana*) was called simply "possum" during the Civil War, and still is in many parts of its range. Virginia opossum, the scientifically recognized common name, is derived from the Algonquian Indian word *apasum* for the animal and the state where it was first described. Opossums are found throughout most of the eastern United States and Central America. They are unique in being the only native marsupial in the United States. The young of marsupials are born incompletely formed and develop for an extended period in the female's specialized pouch (called a marsupium). Adult opossums are about the size of a large domestic cat with a bare tail, and usually long grayish-white fur that is sometimes tanned for leather. Opossums are omnivorous and will eat almost anything, dead or alive. They are especially fond of persimmons, leading to the nineteenth-century term "possum beer" for homebrew made from persimmons.[1] Folklore pertaining to opossums abounds, including the myth that they feign death. During the Civil War, opossums were almost always mentioned in the context of food. Many considered them a delicacy.

Sergeant George W. Bolton, 12th Louisiana Volunteers, in a letter to his family from near Grenada, Mississippi, on Dec. 18, 1862: "The boys in my mess got about twenty pounds of butter, a ham of bacon, besides a good deal of cakes, and Jim Sutton got a baked opossum which was splendid indeed."[2]

Colonel Thomas W. Higginson, 1st South Carolina [African American] Volunteers, near Beaufort, South Carolina, on Nov. 24, 1863: "today I dined on roasted opossum. Done to perfection, done brown, with such cracklings as Charles Lamb in his visions of roast pig only dreamed of, I found it a dish of barbaric fascination. Bear meat is delicious, it is like beef that has been fed on honey; alligator steaks are a kind of racier fried-halibut; but I see that 'possum is one of the great compensations of Nature."[3]

Lieutenant William R. Montgomery, 1st Georgia Sharpshooters, at Lookout Mountain, Tennessee, on Nov. 2, 1863: "Received my box [from home] today. Had a good time. Had Opossum, Chicken, Potatoes, Butter, peach pies & some oh! such nice apples. Like to hurt myself eating."[4]

Private Melville Follett, 42nd Illinois Infantry, near Stones River, Tennessee, on Feb. 3, 1863: "Bennett Goodrich and myself went hunting Possom. Cut down one tree but were not lucky enough to find one."[5]

Colonel Thomas W. Higginson, 1st South Carolina [African American] Volunteers, near Beaufort, South Carolina, on Dec. 1, 1862: "I reproached one [soldier] whom I found sitting up by a camp fire cooking a surreptitious opossum—telling him he ought to be in bed after his hard work."[6]

Private S. O. Bereman, 4th Iowa Cavalry, near Helena, Arkansas, on Feb. 4, 1863: "Our Co. on picket. Snowed all day & at night set in to rain-ing, and freezing. What a dismal night to be out in. In the morning killed several rabbits. One of the boys 'caught a Tarter' in the shape of an opossum. He did not catch it either, but bought it of a darkie who had, and who told him it would be good to eat. The soldier—who had never seen one before—brought home his prize and gave it to our Boy Jake to cook but Jake soon discovered that it was not eatable, as it had nine little ones hidden in its pockets! I have heard of people getting fooled by opossums before— but not just in that way. It was the first I ever saw."[7]

Opossums as a commodity were not exempt from the laws of supply and demand, as is reflected in the following example of 300 percent inflation in less than a year.

John B. Jones, clerk to the Confederate secretary of war, in Richmond, Virginia, on Jan. 18, 1863: "Common tallow candles are $1.25 per pound; soap, $1.00; hams, $1.00; opossum $3.00."[8]

John B. Jones, clerk to the Confederate secretary of war, in Richmond, Virginia, on Nov. 21, 1863: "I saw to-day, suspended from a window, an oppossum dressed for cooking, with a card in its mouth, marked 'price, $10.' It weighed about four pounds."[9]

OYSTERS

During the Civil War, oysters (family *Ostreidae*) were so abundant in Chesapeake Bay that they may have filtered that estuary's entire water volume in less than a week, a process that would now take a year. Oysters were common throughout coastal regions of the Atlantic seaboard and Gulf of Mexico before pollution in the form of toxins and excess nutrients eliminated many natural reefs. As a group of mollusks highly prized for human consumption, their role in maintaining good water quality in marine ecosystems was poorly understood in the nineteenth century, leading to overexploitation.

Oyster shells were part of an important construction material for sturdy structures in those coastal areas lacking natural building stone. Tabby, consisting of a mixture of equal volumes of lime, sand, water, and oyster shells, was poured into forms like concrete. Many coastal forts manned during the Civil War were originally built of tabby in the eighteenth and early nineteenth centuries. Examples include Forts Sumter and Palmetto in South Carolina, Fort Macon in North Carolina, Fort Powell in Alabama, and Fort Sabine in Texas. Commanders learned quickly that tabby walls, often effective against solid shot, were no match for rifled cannon with explosive shells.

For thousands of years humans have consumed oysters wherever they are found around the world. Middens, piles of waste oyster shells, remain as evidence in many areas. Civil War participants frequently mention oysters as food items and often refer to them as curiosities or in a gourmet sense, perhaps because a general lack of refrigeration enhanced the value of fresh oysters.

Charles T. Quintard, chaplain, 1st Tennessee Infantry, at Pulaski, Tennessee, on Dec. 22, 1864: "Dined with Major Jones, and for wonder as to time and place, had oyster soup. General Hood and myself enjoyed the rarity."[1]

Private John F. Brobst, 25th Wisconsin Infantry Regiment, near New Bern, North Carolina, on Feb. 21, 1865, in a letter to his future wife: "We have plenty of oysters here. All that we have to do is go down on the beach and pick them up and open the shell, and you have them all nicely dressed. . . . It is quite a curiosity to see an oyster bed. They grow in very large bunches, sometimes two or three dozen in a bunch."[2]

Sergeant George A. Remley, 22nd Iowa Volunteers, in a letter to his brother from Indianola, Texas, on Feb. 16, 1864: "Well, when I got there one of the men got out his oyster knife and just as he opened them I took them in charge and soon disposed of about two dozen of the largest oysters I ever saw."[3]

Confederate nurse Kate Cumming in Mobile, Alabama, on Jan. 5, 1865: "The enemy have deprived us of one great luxury since taking possession of the bay; that is, oysters. They are not to be had, unless at an exorbitant price."[4]

John Hay, assistant secretary to President Lincoln, at Washington, D.C., on Nov. 8, 1864, while awaiting the vote count of the recent presidential election: "Towards midnight we had supper, provided by Eckert. The President went awkwardly and hospitably to work shoveling out the fried oysters. He was most agreeable and genial all the evening in fact."[5]

Sergeant Taylor Peirce, 22nd Iowa Infantry, writing to his wife on Feb. 3, 1864, at Indianola, Texas: "There is plenty of oysters in the bay and we can get them easily if we try but we have become used to them so that they have ceased to be a luxury."[6]

William B. Gould, African American sailor aboard the USS *Niagara* near El Ferrol, Spain, on Feb. 15, 1865: "The Bay is covered all day with Fishermans. We had some large oysters to day the first since we left the states."[7]

Private Theodore F. Upson, 100th Indiana Infantry Volunteers, near Beaufort, South Carolina, on Jan. 18, 1865: "We are just feasting on them [oysters]. There are a kind here called the cluster oyster. The Darkies bring them into camp and sell them for 10 cts a quart. Some of our boys got a boat and went out on the flats after them. They soon got a boat load, but the tide came in and swamped thier boat and drowned two of them."[8]

Assistant Surgeon William Child, 5th New Hampshire Volunteers, in a letter to his wife from Point Lookout, Maryland, on Nov. 17, 1863: "There are plenty of oysters here. We have them fresh from the water when we like. . . . There is one oyster bed about ten rods in the rear of our tents which we intend to operate as soon as we can get a boat."[9]

Sergeant Allen M. Geer, 20th Illinois Volunteers, near Savannah, Georgia, on Jan. 5, 1865: "Had a morning lunch of three dozen roast oysters only 2 dollars per bushel. Some German people here make money cooking oysters for soldiers and in taking them from the shell."[10]

Private Samuel P. McKee, 22nd Georgia Volunteers, in a letter to his family from Portsmouth, Virginia, on Dec. 26, 1861: "me and two other boys slipped off and went to the river, hooked a boat and some grabs and went a oyster hunting. I got about a bushel, but had nothing but my coat pockets to bring them back in. I knocked off the shell and stewed them for my dinner. They are the first I ever caught or cooked."[11]

Assistant Commissary John G. Earnest, 79th Tennessee Infantry, at Mobile Bay, Alabama, on Dec. 3, 1862: "By the way I called as soon as landing and got a first rate dish of oyster soup for breakfast—the same for dinner and supper."[12]

Then as now, eating raw oysters for the first time was best not undertaken while in a reflective state of mind.

Private Nelson Stauffer, 63rd Illinois Infantry, near Savannah, Georgia, on Dec. 29, 1864: "had plenty of oysters but not much of anything else. Here I first learned to eat them from the shell. I had to or go hungry I opened one and the longer I looked at it, the more decided I was not to try it. I threw it away, went back to quarters and layed around until I got hungrier I opened another but couldn't make it the boys got to making fun of me so I opened another. Shut my eyes, put it in my mouth but a strange feeling came up in my throat and pushed it out. After it was out I noticed that it did not taste badly and after a few moments I rather craved the taste and tried it again, this time it went down and while it did not taste badly, it was some time before I was right sure it was going to stay."[13]

⟵ *Oysters* was also a term for a southern dish made of eggs, butter, and coarse cornmeal but lacking that invertebrate.[14]

RABBITS

The eastern cottontail rabbit (*Sylvilagus floridanus*) was one species of wild mammal that actually benefited from the throngs of forest-clearing pioneers. Although found throughout a range of habitats in eastern North America, fields, gardens, and other clearings surrounded by brushy fencerows and hedges provided ideal homes for an animal naturally adapted to thrive on the forest edge. In these areas during the Civil War, cottontails were obviously abundant.

Ecologically, rabbits are important components of natural food webs by serving as prey for a host of mammalian and avian predators. Humans too have long pursued rabbits for food. However, even when subject to intense hunting pressure by humans and other predators, rabbit populations are normally sustainable because of their high reproductive potential. Females are capable of producing eight litters per year. Civil War participants sought rabbits opportunistically, as any source of protein was relished in lean times.

Lieutenant William R. Montgomery, 1st Georgia Sharpshooters, near Greenville, Tennessee, on March 22, 1864: "We are actually suffering for something to eat. It is still snowing. Heaviest snow I ever saw in my life. Boys are expecting a good time catching rabbits tomorrow." **On the following day he writes:** "The snow is from 15 to 18 inches deep. Have a beautiful day. Boys are all out catching rabbits."[1]

Assistant Commissary John G. Earnest, 79th Tennessee Infantry, near Vicksburg, Mississippi, on March 12, 1863: "In the afternoon, Captain Neale, Lt. Martin and I went a hunting. I killed a squirrel and then we got a rabbit up and had quite a nice chase. I got a shot at him but was so nervous that I missed him. However Dan and the rest of the pack did better for they strung him up. We then returned to speculate on the prospect of Rabbit soup."[2]

Sergeant John Westervelt, 1st New York Volunteer Engineer Corps, camped near the Appomattox River, Virginia, on Feb. 16, 1865: "Rabbits are quite plenty every time it rains it drives them out of their low ground and ravines. If it were not for so many running through the woods and robbing them I would set traps by means I could catch a great many. This morning some darkies and a dog caught two near our cabin."[3]

Sergeant John Q. A. Campbell, 5th Iowa Infantry, in Boonville, Missouri, on Sunday, Dec. 29, 1861: "Wrote a long letter to mother. Rabbit for dinner."[4]

Captain Theodore A. Dodge, 101st New York Volunteers, near Centerville, Virginia, on Dec. 4, 1862: "On the way back we supped at a Mrs.

Robertson's, a secessionist's, off chicken and wild rabbit. I hope it was not poisoned."[5]

Private William G. Bentley, 104th Ohio Volunteer Infantry, in a letter to his family from Frankfort, Kentucky, on Jan. 14, 1863: "We have had several treats lately. One of the boys bought a rabbit pie and an enormous roast turkey which we enjoyed amazingly."[6]

Sergeant Edwin H. Fay, Minden [Louisiana] Rangers, near Grand Junction, Tennessee, in a letter to his wife on April 21, 1862: "I believe I wrote you that I had bought a Navy repeater for $45. I go out and kill rabbits with it almost as well as if I had a rifle."[7]

Soldiers also enjoyed the recreational aspects of pursuing rabbits.

Private Theodore F. Upson, 100th Indiana Infantry Volunteers, at Grand Junction, Mississippi, on Jan. 22, 1863: "We were coming in off pickett this morning and the Brigade was out having drill without arms. They were in line when a rabbit jumped up in front of them. One of the men thoughtlessly tried to catch it and a dozen or more tried to help him. They forgot all disciplin and took after that rabbit. Theier officers shouted at them but no use. The poor rabbit ran this way [and] that between thier legs and under thier feet and finaly ran in his hole. It was too funny for any use. When Colonel Williams got them back in line he was laughing so he could not scold any and had to let it go at that. One man however was badly hurt and was sent to the hospital."[8]

Captain John Dooley, 1st Virginia Infantry, in camp near Winchester, Virginia, in the autumn of 1862: "At times there is a tremendous sudden wild and continued shouting, the whole bivouack is in confusion and wild commotion. Finally the cause is known, and a poor old hare, frightened to death, or captured by the endless bands of pursuers, is dangled through the camp by his hind legs in the hands of his capturers."[9]

Other, less common species of rabbits were found in the Civil War theater. Marsh rabbits (*Sylvilagus palustris*) inhabit coastal regions of the eastern seaboard. Swamp rabbits (*S. aquaticus*) flourish in the forested wetlands of the lower Mississippi River Valley. The following account likely refers to the black-tailed jackrabbit (*Lepus californicus*), a western member of the rabbit family.

Sergeant Taylor Peirce, 22nd Iowa Infantry, writing to his wife on Feb. 3, 1864, at Indianola, Texas: "There is also the Large Rabit or hare. We see them in the mornings playing around. They look almost as large as a fawn and are very good eating."[10]

Various kinds of meat from the spoils of foraging expeditions, sometimes unauthorized, were often referred to as "rabbit."

Private Wilbur Fisk, 2nd Vermont Volunteers, near Harpers Ferry, West Virginia, in a letter to his hometown newspaper on Aug. 23, 1864: "Often [during foraging expeditions] we run across a good fat rabbit, and whether its dress is wool [as from sheep] or bristles [as from hogs] it is all the same to us."[11]

RATS

Rats and mice can be divided into two groups—those native to the New World and those found originally in the Old World. New World examples include the white-footed mouse (*Peromyscus leucopus*) and wood rats (*Neotoma* spp.). Old World examples include the house mouse (*Mus musculus*), black rat (*Rattus rattus*), and Norway rat (*R. norvegicus*). Now found throughout the Americas, the Eastern Hemisphere rodents stole a ride with Europeans wherever they traveled. Black rats are thought to have reached North America in the sixteenth century, with Norway rats following in the mid-eighteenth century. Today, Norway rats have replaced black rats in most of the United States.[1]

During the Civil War the highest-profile rodents were Norway and black rats. Adapted to living in close proximity to humans, these species are infamous for their abilities to consume vast quantities of human foods such as grains, and for their propensity for carrying diseases communicable to man and his domestic animals. Diseases transmitted by these rats or their parasites include bubonic plague, trichinosis, typhoid fever, diphtheria, tularemia, and rabies.[2]

Civil War literature often mentions one sensational aspect of the relationship between rats and humans—the desperate situations when humans were forced to seek rats as food. During extended sieges such as that at Vicksburg, starving soldiers ate rats to survive.[3] Sometimes prisoners of war were likewise placed in such dire conditions.

Captain John Dooley, 1st Virginia Infantry, as a prisoner of war on Johnson's Island in Lake Erie, Ohio, in September 1864: "Rats are found to be very good for food, and every night many are captured and slain. So pressing is the want of food that nearly all who can have gone into the rat business, either selling these horrid animals or killing them and eating them. There are numbers in the drains and under the houses and they are so tame that they hardly think it worth while to get out of our way when we meet them."[4]

Lieutenant Edmund D. Patterson, 9th Alabama Infantry, as a prisoner of war on Johnson's Island in Lake Erie, Ohio, on Sept. 17, 1864: "For several days some of the boys have been killing and eating rats, of which there are thousands in the prison. I have often been hungry all day long, indeed so hungry that I felt sick, and still I could not screw my courage up to the point of eating rats. But today after getting a few mouthfuls of beef and bread, and having been hard at work most of the day on kitchen detail I was constrained to try a mess of rats. My friend, Jones, had been very lucky and had captured a sufficient number of rats to make a big stew and invited me to try them, and it would have done a hungry man's soul good just to have seen me eat them. I cannot say that I am particularly fond of them, but rather than go hungry I will eat them when I can get them, though they have become the fashion to such an extent that from twenty five to a hundred are killed every night at each Block and they are already getting scarce. They taste very much like a young squirrel and would be good enough if called by any other name."[5]

At times soldiers in the field also considered rats table fare, and with noncombatants were often subjected to the general discomfort and anxiety caused by rats.

Lieutenant Rufus Kinsley, 74th United States [African American] Infantry, in a letter to his sister from Cat Island, Mississippi, on Jan. 22, 1865: "We have been short of rations some of the time, and when unable to catch fish, oysters and alligators, on account of the weather, have been obliged to resort to rats, and acorns, both of which are found here in great abundance. The rats live mostly on acorns, and are very good."[6]

Sergeant Bartlett Yancey Malone, 6th North Carolina Infantry, at Point Lookout Prison, Maryland, on Jan. 1, 1864: "I spent the first day of January 64 at Point Lookout, M.D. The morning was plesant but toward eavning the air changed and the nite was very coal. was so coal that five of our men froze to death befour morning. We all suffered a great deal with coal and hunger too of our men was so hungry to day that they caught a Rat and cooked him and eat it."[7]

Sergeant Rice Bull, 123rd New York Volunteer Infantry, at Murfreesboro, Tennessee, on Oct. 14, 1863: "When we reached there at ten that night we were taken to the Court House for quarters. This building had been used by both the Confederate and Union Armies, and I do not think it had been cleaned since the war started. To say it was dirty would have been praise; it was overrun with rats and vermin and unbearably filthy. The rats seemed to live on the food the soldiers threw or dropped on the floor when they were eating. They were almost as tame as cats and ran in every direction under our feet. They had holes in the sides of the room and would dive in them like a flash if you made after them. . . . Tired as we could be, we spread our blankets and tried to sleep but that was impossible. The rats were all over us and they squealed and fought."[8]

Captain William J. Bolton, 51st Pennsylvania Volunteers, near Fredericksburg, Virginia, on Aug. 4, 1862: "Our trip had been a pleasant one, and nothing of particular note occurred except that the boys declare and swear that they had been drinking rat coffee during their trip. After it was too late it appears that a large cauldron had been filled with ship's water during the night to have ready for the next morning. During the night it was discovered that two rats had in some way fallen into the cauldron, and unable to get out had drowned. It remained a standing joke for some time with the boys."[9]

Confederate sympathizer Myra Inman at Cleveland, Tennessee, on July 5, 1864: "I spent a miserable night last night. A rat got in Sister's and my bed in the night."[10]

Charles T. Quintard, chaplain, 1st Tennessee Infantry, at Okolona, Mississippi, on Nov. 15, 1864: "Ye Rats! Ye Rats! For size and multitude the Okolona rats cannot be excelled. All the night long they played the most fantastic tricks in the room we occupied. Once I got up and lighted the fire to drive them off."[11]

John B. Jones, clerk to the Confederate secretary of war, in Richmond, Virginia, on Feb. 11, 1863: "Some idea may be formed of the scarcity of food in this city from the fact that, while my youngest daughter was in the kitchen to-day, a young rat came out of its hole and seemed to beg for something to eat; she held out some bread, which it ate from her hand, and seemed grateful. Several others soon appeared, and were as tame as kittens. Perhaps we shall have to eat them!"[12]

Eliza W. Howland, Union nurse, in a letter to her husband from Washington, D.C., on Jan. 28, 1862: "Mr. Hopkins told us of one poor fellow of a Vermont regiment who was brought to the hospital in Alexandria with typhoid fever, having both feet frozen and one of them eaten by rats!"[13]

SNAKES

Despite stories that the first fatality of the Civil War was caused from a coral snake bite, no reliable documentary evidence of the claim has surfaced. One popular magazine writer made the following relevant comments in 1910: "Nowhere in the Civil War records does a death from this cause [snakebite] appear, though hundreds of thousands of men were living 'on the country,' and at a time when the serpent clan was much more numerous than now."[1] In that era the outlandish myths that surrounded snakes were more prevalent than today, and the irrational fear of these reptiles was just as common. Hoop snakes, stinging snakes, snakes that whip people to death, and snakes that steal milk from cows were thought to exist and threaten all humanity since the days of the Garden of Eden. Other than king snakes (*Lampropeltis* spp.), which were barely tolerated because they sometimes feed on other snakes, snakes were considered dangerous vermin and eradicated opportunistically. Then as now, it is likely that more people injured themselves trying to kill harmless snakes than were harmed by poisonous ones. However, in a time before the development of antivenin, bites from poisonous snakes likely had higher mortality rates than those treated with modern therapies. As a result, a host of home remedies surfaced, most of which were ineffectual. One involved the immediate decapitation of a black hen into whose warm innards the bitten hand or foot was abruptly plunged.[2] Following the pattern of most Civil War medicines, those used to treat snakebite contained a high percentage of alcohol.

About forty-seven species of snakes occur in the eastern United States, seven of which are poisonous. As a group they are highly adaptive and found in most habitat types. Snakes are important, unappreciated components in many ecosystems and as predators often aid in the control of other animal populations such as rodents. The loss of habitat is the greatest threat to snakes today. Civil War participants had countless encounters with snakes. Their writings reflect a general misunderstanding of the reptiles, not unlike the situation today.

Rattlesnake (*Crotalus* spp.) encounters seemed to be especially noteworthy.

Sergeant Edwin H. Fay, Minden [Louisiana] Rangers, near Priceville, Mississippi, in a letter to his wife on June 10, 1862: "We came out some ten miles from Houston [Miss.] that night and I slept with a rattlesnake all night. Discovered him next morning and killed him, so this time I was preserved from danger."[3]

Private Theodore F. Upson, 100th Indiana Infantry Volunteers, near Vicksburg, Mississippi, on July 27, 1863: "I heard a racket yesterday and went to see what it was all about. I found that some of the boys had captured one of the big yellow rattle snakes that are found here and had built a pen of sticks at the foot of a white oak tree that had a limb sticking out over the pen and snake. When they had got it all fixed a man from the 8th Wisconsin came bringing the eagle they carry instead of a Regimental flag. The man carried it on sort of a hod with a flat top shaped like a shield with a perch on it. When he came the boys were betting on the result of the encounter betwen the eagle and the snake. Some thought the eagle would not tackle the snake, some that if he did the snake would bite him. The boys had teased the snake with sticks till he would strike at any thing that moved, and do it quick too. The man with the eagle (he is named Old Abe) was taking all bets that were offered saying the eagle will kill the snake. . . . Finaly the carrier gave Old Abe a little toss and he flew up on the limb where he sat turning his head first to one side and then the other, looking down at the angry rattler below. Then his keeper said, 'Take him Abe.' And before I could see how it was done he gave a scream, droped from the limb, and with one claw seized the rattler by [t]he head, and with the other on his body literaly tore his head off, then hopped up on the limb again. I would have lost my money sure. The rattler had no chance to bite."[4]

Refugee Kate Stone in Lamar County, Texas, on Aug. 30, 1863: "we just missed driving over the largest rattlesnake, stretched across the road basking in the sun. It was larger than my arm and had twelve rattles. That frightened us most of all. It might have glided into the carriage as we drove over it."[5]

John M. Follett, 33rd Illinois Volunteer Infantry, at Pitman's Ferry, Arkansas, in a letter to his wife on April 25, 1862: "Enclosed I send you the drawing of the rattlers of a snake killed by one of our boys. It is just the size and number."[6]

Captain Charles B. Haydon, 2nd Michigan Infantry, near Vicksburg, Mississippi, on July 5, 1863: "Water scarce. Killed a very large rattlesnake in camp."[7]

British Lieutenant Colonel Thomas Freemantle at Lake Concordia, Louisiana, on May 14, 1863: "These bayous and swamps abound with alligators and snakes of the most venomous description. I saw many of the latter swimming about exposed to a heavy fire of six-shooters; but the alligators were frightened away by the leading boat."[8]

Colonel Thomas W. Higginson, 1st South Carolina [African American] Volunteers, at Port Royal Island, South Carolina, on May 2, 1863: "The snake [brought in by a soldier] above named was called the thundersnake by the men, of which I could get no explanation save 'speck he look like streak lightnin, sa.'"[9]

Corporal Rufus Kinsley, 8th Vermont Regiment, on Ship Island, Mississippi, on April 11, 1862: "Went to the east end. Saw a huge anaconda hanging over the limb of a tree. Left the vicinity on the 'double quick.' Helped kill an alligator."[10]

Private Gottfried Rentschler, 6th Kentucky Volunteer Infantry [Union], in a letter to the *Louisville Anzeiger* from near Acworth, Georgia, on June 10, 1864: "we lay in the evenings on the hard earth with a smile on our lips anticipating the delights of the ensuing sweet dreams about our loved-ones at home; that here in Georgia all too often will certainly be interrupted by the long, black snake, of which there are a large number in the local forests; and which not infrequently crawl over our faces while we are asleep. Snakes are the symbol of slyness. They want to remind us through their presence that the soldier should sleep with his eyes held open?!"[11]

Private Isaac Jackson, 83rd Ohio Volunteer Infantry, at Milliken's Bend, Louisiana, on March 22, 1863: "And snakes, oh, my! I expect they will be so thick that we can hardly walk in a month or two. Next morning two or three of us went along the ditch on which we were standing picket and we killed 5 small snakes and seen two more. You may know that will do for so early in the season."[12]

Captain William J. Bolton, 51st Pennsylvania Volunteers, at Newbern, North Carolina, on June 6, 1862: "The camp is full of green snakes, and the boys are kept busy in killing them."[13]

Mary Vermilion in a letter to her husband in the 36th Iowa Infantry from Iconium, Iowa, on Aug. 10, 1863: "I went into my room a minute ago . . . and almost stepped on quite a large snake lying on the floor. I was so scared (I was always afraid of snakes) that before I could do anything it ran under the bed. . . . I am going to make a raid in there presently. . . . If I don't kill the snake . . . I shall be afraid to sleep in there anymore. I shouldn't fancy waking up in the night and finding a snake in my bed, ugh! Well, I have just had a big hunt for my snake, love, but I couldn't find him anywhere. I guess he escaped outdoors while I was getting the shovel to kill him, and getting over my scare. If he will stay away now, it is all I ask of his snakeship."[14]

Private Alexander Downing, 11th Iowa Infantry, near Monroe, Louisiana, on Aug. 27, 1863: "This is the worst bivouac we have yet occupied. It is full of poisonous reptiles and insects, centipedes, jiggers, woodticks, lizards, scorpions and snakes of all kinds—I have never seen the like. Some of the boys killed two big, spotted yellow snakes and put them across the road—they measured about fifteen feet each.[!]"[15]

↚ Copperheads—Northern Democrats who opposed the Union's war policy and favored a negotiated peace.[16]

SQUIRRELS

Several species of squirrels are found in the eastern United States. Those mentioned most frequently by soldiers were likely gray squirrels (*Sciurus carolinensis*) and fox squirrels (*S. niger*). Both species are common tree squirrels and prefer similar habitat, usually mature mixed hardwood forests with an abundance of acorns and other mast fruits. Gray squirrels tend to be more common in second-growth forests with dense understories, while fox squirrels are prevalent in small woodlots and open pine forests.

Squirrels may have unwittingly contributed to Civil War casualties. Thousands of rural boys learned to handle rifles and develop marksmanship skills in pursuit of wily squirrels that provided abundant, challenging targets. These soldiers brought years of firearms experience and practice to the ranks—a proficiency otherwise not likely attained after enlistment.

When the South began to suffer a dearth of basic necessities, shoes for women and children were sometimes made of squirrel-skin uppers fastened to leather or wooden soles.[1]

Article in the *Meridian Mississippian*, Aug. 29, 1863: "An exchange says that squirrel skins, tacked down to a board, the hair next to the board, with hickory ashes sprinkled over them, for a few days, to facilitate the removal of the hair, and then placed in a strong decoction of red-oak bark, will, at the end of four days, make excellent leather, far stronger and tougher than calf skin. Four skins will make a pair of ladies' shoes. We hear that the ladies of some of the intertor counties are wearing these shoes, and find them equal in softness and superior in durability to any others. The longer the skins are left in the decoction of the bark the better the leather. By this plan anybody may have a tan yard, and make their leather, as the skins are easily and cheaply procured, and any vessel holding a gallon will serve as a vat. Our readers will do well to try it."[2]

People thought of squirrels most frequently during the Civil War in the context of food. Squirrels provided an often-scarce source of protein, especially for southern soldiers and citizenry. Insignificant as food on a large scale, squirrels were nevertheless a welcome ingredient in many campfire mulligans.

Sergeant Edwin H. Fay, Minden [Louisiana] Rangers, near Camden, Arkansas, in a letter to his wife on April 25, 1864: "Tell Bud to take my rifle up to Mabry and have it fixed, the lock repaired and barrel dressed out, rifles recut. If I had it with me I could kill squirrels and we could live a little better. We are living harder here in the front than ever I did before except at Corinth."[3]

Thomas M. Chester, African American correspondent, in a dispatch to the *Philadelphia Press* from near Richmond, Virginia, on Sept. 7, 1864: "There are sure evidences of a scarcity of food just now among the rebels. . . . The men acknowledge that they receive far below the usual quantity allowed to them, in consequence of its great scarcity, and as yesterday was spent by their pickets in gaming for squirrels and other things, to be obtained by shooting, we may very correctly infer that they are in a very hungry condition."[4]

Private S. O. Bereman, 4th Iowa Cavalry, near Vicksburg, Mississippi, on Aug. 28, 1863: "Col. Hammond & I went out squirrel hunting—that is the Col. shot the squirrels & I held his horse! It was too windy and we didn't kill but one."[5]

Private Henry C. Caldwell, 7th Louisiana Infantry, in northern Virginia on Nov. 16, 1861: "I took a hunt while at Dr. Pane [and] killed one squirrel and one partridge."[6]

Assistant Commissary John G. Earnest, 79th Tennessee Infantry, at Jackson, Mississippi, on Jan. 15, 1863: "When I got up I found that a snow of about three inches depth had fallen. After taking a smoking breakfast of biscuit, squirrel, eggs & coffee, we warmed up a little and bid our kind friends adieu."[7]

Confederate sympathizer Myra Inman at Cleveland, Tennessee, on Oct. 9, 1863: "Cousin John came in this eve and brought us two squirrels."[8]

Major James A. Connolly, 123rd Illinois Infantry, writing to his wife on March 20, 1864, from camp in Ringgold, Georgia: "Tomorrow, after getting my small pox [vaccination] started, I shall go out and shoot some squirrels, if it is a pleasant day."[9]

A. L. Peel, Adjutant, 19th Mississippi Regiment, on leave at home near Holly Springs, Mississippi, on Feb. 27, 1863: "Add Volney and I went hunting this morning and killed Ten Squirrels. came home to Dinner Shot at a mark I beat them. We went out this evening & killed Eight Squirrels out of Two trees."[10]

Sergeant John Q. A. Campbell, 5th Iowa Infantry, in Boonville, Missouri, on Jan. 7, 1862: "Squirrel for breakfast—rabbit for dinner."[11]

Colonel Thomas W. Higginson, 1st South Carolina [African American] Volunteers, near Beaufort, South Carolina, on Dec. 5, 1862: "Once William produced with some palpitation something fricasseed which he boldly termed chicken, it was very small & seemed in some undeveloped condition of ante-natal toughness—after the meal he frankly avowed it for squirrel."[12]

Private William M. McKee, 18th Georgia Volunteers, in a letter to his sister from near Richmond, Virginia, in September 1861: "Tel Mandy to take good cear of Triv [a squirrel dog] for I want to shoot some squirrels."[13]

Sergeant Samuel A. Clear, 116th Pennsylvania Volunteers, just returned home to Uniontown, Pennsylvania, after the war, on June 21, 1865:

"Buckwheat Cakes and honey (wild) has to suffer and all else that is eatable, that appetite of mine still continues. Grey Squirrel and Red Squirrel Pot Pie &c &c—Hardtack a thing of the past."[14]

Squirrels also provided sport for bored soldiers. At times men chopped down numerous trees just for the prize of capturing a fleeing squirrel barehanded. Additionally, squirrels as pets were a source of entertainment and diversion from the horrors of war.

Private Robert A. Moore, 17th Mississippi Regiment, near Leesburg, Virginia, on Nov. 24, 1861: "This is Sunday in camp & a number of the boys are amusing themselves after squirrels. They have caught 3 or 4. I never heard such a noise in my life. Cpt. Duff was up a tree with a stick after a squirrel."[15]

Major Silas T. Grisamore, 18th Louisiana Infantry, near Monroe, Louisiana, on Jan. 30, 1864: "The soldiers used to amuse themselves catching squirrels, and if one of those little animals ever showed his head, they got him, certain. Whilst in this camp, one of the little creatures was discovered, and the boys went to work and cut down 15 trees averaging two feet in diameter before they caught him. The amount of hollering and yelling during the time must be left to the vivid imagination of the reader."[16]

Private John Westervelt, 1st New York Volunteer Engineer Corps, camped near the Appomattox River, Virginia, on Feb. 16, 1865: "On Sunday our drivers catched a large gray squirrel, To day I gave him 25 cts for him and I have him alive. I have made a cage for him and he is getting quite tame. I will sell him if I can if not I will try and bring him home when I come."[17]

Lieutenant Colonel Charles F. Johnson, 81st Pennsylvania Volunteer Infantry, in a letter to his wife from Yorktown, Virginia, on May 8, 1862: "bye-the-bye—my two Flying Squirls are dead, having been smothered in their transportation."[18]

Refugee Sarah Morgan Dawson on Linwood Plantation near Port Hudson, Louisiana, on March 10, 1863: "Poor Mr. Halsey! What a sad fate the

pets he procures for me meet! He stopped here just now on his way some-
where, and sent me a curious bundle with a strange story, by Miriam. It
seems he got a little flying-squirrel for me to play with (must know my par-
tiality for pets), and last night, while attempting to tame him, the little
creature bit his finger, whereupon he naturally let him fall on the ground,
(Temper!) which put a period to his existence."[19]

TURTLES

As a unique type of reptile, turtles are characterized by ribs expanded
and modified into a protective shell. All freshwater habitats in the eastern United States are home for turtles. About thirty species of aquatic turtles ranging in size up to the two-hundred-pound alligator snapping turtle (*Macrochelys temminckii*) frequent rivers, streams, lakes, ponds, and
marshes. Two species (eastern box turtle, *Terrapene carolina*, and gopher
tortoise, *Gopherus polyphemus*) are terrestrial, and six species of sea turtles are found along the Atlantic seaboard and Gulf coasts. All turtles lay
eggs in burrows on land, with sea turtle clutches sometimes exceeding one
hundred. Turtle eggs and meat have been valued as food by humans for
thousands of years, leading to the near eradication of some species. The di-
amondback terrapin (*Malaclemys terrapin*), a resident of brackish marshes
and estuaries and now greatly reduced in numbers, was once reported
abundant enough to be a mainstay for feeding slaves in Maryland. Now
banned from commercial use by international treaty, tortoiseshell from
hawksbill sea turtles (*Eretmochelys imbricata*) was used during the Civil
War era to produce combs, jewelry, hair ornaments, picture frames, and
snuff boxes.[1]

**Private John Westervelt, 1st New York Volunteer Engineer Corps, near
Hilton Head, South Carolina, on Feb. 15, 1864:** "Today while taking
a stroll back of the camp I found a terrapin (turtle) weighing about 5 lbs.
This was quite a prize, particularly as fresh meat has been unusually scarce
lately. It was a female and I suppose was about to deposit its eggs in the
sand, which accounts for its being so far from the water. Terrapin is considered one of the most delicious of the turtle species. I do not know how
large they grow, but on the voyage from New York I saw thousands of them
floating on the sea that did not appear to be much larger than this one. I

took from it 15 eggs as large as a small sized hens eggs. I boiled them and found them much richer then hens. . . . I made a stew of the flesh for dinner, and boiled the eggs for breakfast next morning, and they made two of the finest dishes I ever tasted."[2]

Private John Westervelt, 1st New York Volunteer Engineer Corps, on the Appomattox River, Virginia, on June 11, 1864: "Fishing nearly all day and caught a fine lot with the three last hooks sent me. Ed Whittenoure caught a snapping turtle weighing about 15 lbs. As he did not know what to do with him he offered me half if I would cook him so I cleaned and salted him for the night."[3]

Mary Boykin Chesnut at Richmond, Virginia, on Jan. 31, 1864: "Someone sent up a supper of terrapin stew, oysters and Rhine wine, and a box of sugar plums."[4]

Lieutenant Rufus Kinsley, Second Corps d'Afrique, in a letter to his father from Ship Island, Mississippi, on May 29, 1864: "the east end of the island is covered with alligators and turtles. The turtle furnishes excellent meat, and the eggs, which are found in bushels, are quite as good as hen's eggs."[5]

Lieutenant Rufus Kinsley, Second Corps d'Afrique, at Ship Island, Mississippi, on June 15, 1864: "Have had considerable sport hunting turtles, for a month or two; they are found here at this season in great numbers. Most of them weight from three to five hundred pounds each. They come on shore in the night, to deposit their eggs in the sand. A turtle deposits two or three hundred eggs in a night, and leaves them to their fate."[6]

British journalist William H. Russell in a letter to a friend from Washington, D.C., on March 26, 1861: "If the rapidly expanding orb which contains your stout friend's digestive apparatus had been provided with adequate masticatory apparatus & ducts he could have dined seven times a day—terapin soup, canvas backed ducks clam chowder I abhor ye."[7]

British journalist William H. Russell at New Orleans, Louisiana, on May 24, 1861: "Terrapin soup excellent, fish bouillabaisse very good, soft crab, pompinoe &c. wines capital very good service, French cooking—

glorious sunset over the waters of Ponchartrain—yachts & fishers stealing about gardens full of rose laurels in full bloom & the most wonderful
sand flies."[8]

↞ "Pook turtles" was an epithet given to a class of seven gunboats designed
for the Union by Samuel M. Pook. The boats' lack of speed resulted in
the nickname. A restored survivor of the Mississippi River campaign, the
ironclad *Cairo* is on display at the Vicksburg National Military Park.[9]

MISCELLANEOUS INVERTEBRATES

Invertebrates are animals without vertebral columns (backbones). About
98 percent of all animals are invertebrates—every group except mammals,
birds, fish, reptiles, and amphibians. During the Civil War, invertebrates
played sundry roles as human food (oysters), vector of disease (mosquitoes), medicinal tool (leeches), and crop pollinator (bees).

Insects

Insects as a group are the most abundant and diverse creatures on the
planet. The million known species likely represent only a small fraction of
the total kinds present. Nearly every ecosystem on earth harbors insects.
Other than those species discussed above (usually in terms of disease or
misery they caused), insects were mentioned by Civil War participants as
novelties and as figures of speech.

**Private John F. Brobst, 25th Wisconsin Infantry Regiment, near Dan-
forth, Georgia, on May 23, 1864, in a letter to his future wife:** "Boys [of
this regiment are] cursing old Jeff Davis, wanting him tied up to a fence
post and let the grasshoppers to kick him to death. Very good way for him
to die I think, don't you. . . . We are on our way to Atlanta."[1]

**Captain Samuel T. Foster, 24th Texas Cavalry (dismounted), at the bat-
tle of Pickett's Mill, Georgia, on May 27, 1864:** "Our men have no protection, but they are lying flat on the ground, and shooting as fast as they
can. This continues until dark when it gradually stops, until it is very dark,
when every thing is very still, so still that the chirp of a cricket could be

heard 100 feet away—all hands lying perfectly still, and the enemy not more than 40 feet in front of us."[2]

Sergeant Hamlin A. Coe, 19th Michigan Volunteer Infantry, near Atlanta, Georgia, on Aug. 20, 1864: "Last night was the gloomiest of my experiences in this campaign. . . . The night air was made thrice hideous by the roar of battle upon the extreme right, the squeaking of a thousand crickets, and forest insects and the hoot of the owl."[3]

Confederate sympathizer Myra Inman at Cleveland, Tennessee, on Aug. 15, 1863: "A cold, cloudy day very much like fall, the fall crickets have been singing all eve a song, which I do not like to hear."[4]

Lieutenant Theodore A. Dodge, 101st New York Volunteers, near Harrison's Landing, Virginia, on July 3, 1862: "Such a country for productiveness I never saw. Grasshoppers as big as canaries and moths gigantic in size."[5]

Refugee Kate Stone in Lamar County, Texas, on July 16, 1863: "As we sat on the gallery tonight, gazing across the darkening prairie into the gleaming west, the very air was brilliant with fireflies. The fancy came that they were the eyes of the departed Indians, come to look again on their old hunting grounds."[6]

British journalist William H. Russell at Washington, D.C., on July 5, 1861: "Weary evening rather. Afterwards talk of fire flies anglice lightening bugs."[7]

Sergeant Charles B. Haydon, 2nd Michigan Infantry, near Washington, D.C., on June 17, 1861: "The night was quiet. One sentinel mistook a lightning bug for a dark lantern & fired upon it. Some of them are very much inclined to be scared."[8]

Mary Vermilion in a letter to her husband in the 36th Iowa Infantry from Iconium, Iowa, on May 26, 1863: "The caterpillars have killed the crab-apple grove entirely, I am very sorry about it. They are worse than they were last year."[9]

Crustaceans

Crustaceans are a large group of invertebrates that include shrimp, crabs, lobsters, crayfish, and barnacles. Most are aquatic and go through several developmental stages in their life cycle, often shedding exoskeletons in the process. As they have been since the beginning of maritime history, barnacles were a factor to be dealt with by shipping during the Civil War era. Warships, privateers, and other commercial vessels were unable to operate at their potential when burdened with loads of encrusted barnacles. Writers of the period, however, mentioned crustaceans more often in their roles as food for humans. Blue crabs (*Callinectes sapidus*) were found in all coastal areas and were sought after then as today. The avid pursuit of crayfish (crawfish) for the table had not yet escaped the bounds of Louisiana.

British Lieutenant Colonel Thomas Freemantle at Shreveport, Louisiana, on May 8, 1863: "I called on General Kirby Smith, who commands the whole country on this side of the Mississippi. . . . [Gen. Smith's wife] proposed that we should go down to the river and fish for cray-fish. We did so, and were most successful, the General displaying much energy on the occasion."[10]

Private George A. Remley, 22nd Iowa Volunteers, in a letter to his mother from Berwick City, Louisiana, on Sept. 26, 1863: "Stationed along the bank may be seen numerous individuals with pole and line in hand patiently watching and waiting. These are the crab fishers and their 'name' is legion. The sea crab looks something like a craw-fish only more so. The flesh is tender and very delicately flavored. They require no preparation for the kettle, but are thrown into the boiling water while alive and kicking."[11]

Captain William D. Dixon, [Savannah] Republican Blues, at Fort McAllister, Georgia, on May 16, 1864: "I went 6 miles down the river fishing this morning but fish did not bite. I caught a lot of crabs."[12]

Captain William J. Bolton, 51st Pennsylvania Volunteers, at Newport News, Virginia, on July 16, 1862: "The weather very warm, and all spare time is spent in fishing for crabs, bathing, foraging, as our rations are short and very scarce."[13]

Others

Leeches are classified as a group of segmented worms that live in terrestrial, freshwater, and marine environments worldwide. Infamous for their habit of feeding on blood, leeches were a common component of Civil War medicine. Jars or tin boxes containing live leeches were often found in the medicine chests of field physicians in an era before germ theory was widely accepted. Diseases were thought to be a result of an imbalance of the four bodily fluid types, or "humors," blood being one of the critical four. To adjust the perceived imbalance, leeches were applied to a patient to remove blood thought to be causing the disparity. Leeches were of no more benefit to the patients than to the soldiers who became unwilling hosts of the parasite while swimming or bathing.

A. L. Peel, Adjuntant, 19th Mississippi Regiment, near Bull Run, Virginia, on Aug. 10, 1861: "went to the Creek & bathed, the damn Leaches bother a fellow mightily."[14]

Sponges are primitive aquatic animals that depend on water flowing through their porous bodies for oxygen and nutrition. Most species are found in saltwater habitats and are attached to the sea floor or other substrate. For thousands of years humans have used the soft fibrous skeletons of some species for padding or cleaning tools. Such commercial use of sponges has declined significantly as a result of overfishing, and synthetic materials have replaced them today. In the Civil War, medicinal sponges were employed during treatment of injured and sick patients, and their contamination likely contributed to the rampant infections in an era before the value of antiseptic conditions was appreciated. One Confederate surgeon wrote after the war: "One blessing we enjoyed, due to the blockade, was the absence of sponges, clean rags being substituted for them with telling advantage. These rags could be washed, as was done and used over and over again. It is next to impossible, easily, if possible at all, to wash an infected sponge."[15]

> Bearing the bandages, water and sponge,
> Straight and swift to my wounded I go,
> Where they lie on the ground, after the battle brought in;
> Where their priceless blood reddens the grass, the ground. . . .
> —Walt Whitman, "The Wound-Dresser"

❧

Soldiers far from home often pondered the novelty of unusual inverte-brates, from spiders to bizarre sea creatures.

Sergeant George A. Remley, 22nd Iowa Volunteers, in a letter to his fa-ther from Matagorda Island, Texas, on March 24, 1864: "You have fre-quently heard of the tarantula I suppose. They are plenty down here and ugly looking customers. I saw one the other day and I assure you that I don't desire a very intimate acquaintance with any thing of the kind. They look like a huge, fat spider from one to three or four inches across, covered with short, brownish hairs. They have teeth, shaped very much like a go-pher's claws, and with these they bite savagely. Their bite is said to be very poisonous."[16]

Private John Westervelt, 1st New York Volunteer Engineer Corps, at Folly Island, South Carolina, on June 12, 1863: "I must tell you of a crea-ture I found this morning called by sailors a portuguese man of war. It is bright sky blue in colour. It was (probably) brought ashore during a squall last night. . . . It is a verry beautiful creature . . . part of his body is transpar-ent and underneath he is a beautiful crimson and pink."[17]

MISCELLANEOUS MAMMALS

White-tailed Deer

The white-tailed deer (*Odocoileus virginianus*) is the deer of the eastern United States. Once called the Virginia deer, the species is more common today than during the Civil War era. At that time, habitat modification and large-scale, unregulated commercial hunting had initiated a population plunge that ended in the early twentieth century, when the total number of white-tailed deer in the United States was estimated at 500,000. Intensive management, including restrictive harvest regulations and research into the life history of deer, has resulted in a dramatic recovery of the species to a current population of about thirty million.[1]

White-tailed deer were a vital part of many Native American cultures east of the Mississippi River. As providers of food, leather, and a host of other items used in daily life, they played much the same role as bison

(*Bison bison*) did for western tribes. To a lesser extent, deer were also important to many European pioneers in America. Deer or their products played no consequential role in the Civil War. Occasionally they were shot for home consumption or the market when available, and buckskin-clad soldiers were scattered through the ranks.

Kate Stone, Brokenburn Plantation near Milliken's Bend, Louisiana, on Oct. 24, 1861: "Brother [Lt. William R. Stone, Jeff Davis Guards] and many others went hunting early this morning, the first frost of the season whitening the grass, but not enough even to kill the cotton. Jimmy killed the deer, his first victim after so many trials. Johnny and I followed the dogs for some distance. The chase is certainly exciting sport. No wonder men like it so."[2]

Sergeant William D. Dixon, [Savannah] Republican Blues, St. Catherine's Island, Georgia, on Aug. 28, 1861: "All the Commissioned Officers except the Capt went out Deer Hunting this morning with the Overseer of the Plantation."[3]

John B. Jones, clerk to the Confederate secretary of war, in Richmond, Virginia, on Jan. 25, 1864: "I noticed, to-day, eight slaughtered deer in one shop; and they are seen hanging at the doors in every street. The price is $3 per pound."[4]

Private George A. Remley, 22nd Iowa Volunteers, in a letter to his father from Rolla, Missouri, on Dec. 3, 1862: "Company D returned last Monday evening bringing into camp three deer and twelve wild turkeys that they had killed on their way back. They report game very plenty."[5]

❧ "Bucktails"—the nickname of a regiment recruited from the Bucktail region of Pennsylvania. Men in the outfit wore deer tails on their hats.[6]

Raccoons and Other Furbearers

The familiar raccoon (*Procyon lotor*) is the only member of its taxonomic family found in the eastern United States. Raccoons eat a wide variety of plant and animal foods, including wild fruits, nuts, insects, reptiles, crustaceans, fish, and birds. During the Civil War era, they in turn were relished as food by humans, especially in the South. Their omnivorous feeding habits, which included raiding cornfields and gardens as well as chicken houses, resulted in an antagonistic relationship with rural farmers; raccoons were shot or trapped opportunistically. Their pelts were often used in the fur industry. A company in the 6th Alabama Regiment was known as the "Raccoon Roughs," and many of its members wore coonskin caps.[7]

Lieutenant Rufus Kinsley, 74th United States Colored Infantry, in a letter to his brother from Ship Island, Mississippi, on Dec. 11, 1864: "You have had 'Thanksgiving Day' and so have we, since I wrote you last. The Col. sent means for us to have a good thankful time, and the then recent news contributed more largely still; and, with roast coon, alligator stew, oysters, fish, crabs, duck, in the shape of chicken pie, and speeches and music, we had a very pleasant time."[8]

A. L. Peel, Adjutant, 19th Mississippi Regiment, near Orange, Virginia, on Sept. 29, 1863: "The Regt were all after a Coon this morning Caught it, had a race after it."[9]

Sergeant Sidney Carter, 14th South Carolina Volunteers, in a letter to his wife from Pocotaligo, South Carolina, on Jan. 30, 1862: "We had just got back from a picket tour of twelve miles measured with a coon skin."[10]

Both red (*Vupes vulpes*) and gray (*Urocyon cinereoargenteus*) foxes are found in the Civil War arena and were valued for their pelts and recreational use. The sport of fox hunting with hounds was imported from England and was a favorite pastime on many southern plantations. Red fox were the preferred object of chase, as gray fox tend to climb trees when pursued and shorten the hunt. Both species were reviled for their predation on domestic fowl, even though their consumption of vast numbers of rodents provided a beneficial service to farmers that far outweighed their effect on poultry.

Private Robert A. Moore, 17th Mississippi Regiment, at Culpeper, Virginia, on Aug. 5, 1861: "I went out to Mr. Hudson's this evening . . . The old man has a fine pack of fox hounds & says he can have a chase anytime by going three miles. The foxes are all red nearly, & are sometimes caught in one or two hours."[11]

Corporal Robert A. Moore, 17th Mississippi Regiment, near Brucetown, Virginia, on Oct. 24, 1861: "Have been busily engaged erecting our bivouacs. Jumped a red fox in the brigade & the boys caught it."[12]

Beavers (*Castor canadensis*) are large, flat-tailed, dam-building rodents that were once found along most American streams and rivers. Valued for their pelts, North American beavers were the source of immense wealth for European fur dynasties. Easily trapped, beavers were nearly eliminated east of the Mississippi River by the time of the Civil War. In recent years beavers have repopulated much of their historical range, and human/beaver conflicts are common because of the tree damage and flooding associated with their habits. Contemporary beaver problems are now reported from places such as the Richmond National Battlefield Park, where beavers had long been extirpated by the time of the war. President Lincoln often wore a beaver-felt top hat on formal occasions and was doing so the evening he was assassinated.[13]

Lieutenant John Q. A. Campbell, 5th Iowa Infantry, near Vicksburg, Mississippi, on June 26, 1863: "This afternoon I visited a beaver dam on the creek a short distance above the camp and saw some of the ingenious workmanship of those animals."[14]

⤆ The Battle of Beaver Dam Creek, also known as the Battle of Mechanicsville, occurred in Hanover County, Virginia, on June 26, 1862.

The woodchuck or groundhog (*Marmota monax*) is a medium-sized burrowing rodent kin to squirrels. It is found in Canada and eastern North America, except in the Deep South. Woodchuck habitat includes forest edges, meadows, and other openings. The woodchuck population likely increased when settlers cleared heavily timbered areas for farms and crops. Woodchucks play important ecosystem roles as their burrows provide homes for rabbits, foxes, skunks, and other animals. Their extensive burrowing also aerates large volumes of soil over time. Woodchucks were once trapped or shot for their fur, which was used in low-quality garments, and for leather used in straps and laces. During the Civil War, they were considered nuisances for their forays into family gardens and for the crippling hazards to horses caused by their burrows.[15]

Lieutenant Sidney Carter, 14th South Carolina Volunteers, in a letter to his wife from near Berryville, Virginia, on April 23, 1863: "I must tell you of a groundhog that I saw last Sunday dug from his hole and given to me. I ate his meat and on Monday, I dressed his skin to make shoe strings. I will enclose you a pair and a pair for Father."[16]

Sergeant Samuel A. Clear, 116th Pennsylvania Volunteers, near Petersburg, Virginia, on Feb. 2, 1865: "This is ground hog day and he can see his shadow without any trouble. . . . All quiet in front."[17]

AFTERWORD

Impacts of the Civil War on Flora and Fauna

The impacts of the Civil War on flora and fauna are difficult to determine, although from a population perspective nearly all were likely short term. As mentioned elsewhere in this book, the conflict occurred for the most part in a region already greatly altered by human activities. For example, the virgin forests in the area where the Battle of the Wilderness was fought had long since gone to fuel the fires of Catherine's Furnace, an old pig iron smelter.[1] Yet without a doubt, the Civil War caused major landscape alterations on a local scale. The increase in war-related industries and transportation created a demand for wood unequaled in the history of North America. Charcoal made from various hardwood species was critical to iron production. Estimates of the number of acres of woodlands required to make 1,000 tons of pig iron varies from 150 to 1,500, depending on the type of wood and the efficiency of the furnace.[2] Wood was the most common fuel for steamboats during the war, and large ones consumed 50 to 75 cords a day.[3] A round trip from Louisville to New Orleans required 529 cords, by one estimate.[4] Most locomotives used wood instead of coal until after the war. In 1859 about 3,600 of the 4,000 locomotives in the United States were wood fired.[5] Demand for wood for other uses such as cooking, heating, and construction was unparalleled and inestimable. The obvious results were altered forests.

Sergeant Alexander Downing, 11th Iowa Infantry, near Vicksburg on Jan. 27, 1863: "Wood for fuel is becoming very scarce in camp, and also on the transports. The 13th Iowa, with thirty of us from my regiment, were detailed to go with the transports up the river for wood. . . . There are six thousand cords of wood piled up here."[6]

Captain Theodore A. Dodge, 101st New York Volunteers, near Stafford, Virginia, on March 18, 1863: "It is wonderful how the whole country round here is literally stripped of its timber. Woods which, when we came here, were so thick that we could not get through them any way are now entirely cleared—the pine being used for building and making roads, and the cedar and hard wood, of which there is a great quantity, for fire wood."[7]

Plants other than trees were also depleted in some areas, especially those used in the production of pharmaceuticals. Southern laboratories ran newspaper advertisements soliciting specific herbs needed for medical purposes. A surgeon at the Confederate depot in Macon, Georgia, reported in July 1863 that he possessed 16,034 pounds of indigenous remedies ready for service and 64,779 pounds of unprocessed plants.[8] Ship manifests often revealed large cargoes of herbs such as arrowroot,[9] and soldiers sometimes exploited plants directly for practical uses.

Captain William J. Bolton, 51st Pennsylvania Volunteers, at Roanoke Island, North Carolina, on Feb. 26, 1863: "Strolled on further and found the soldiers were digging out in the swamps large quantities of brier roots [*Smilax* spp.] in which the island abounds, and in their leisure time convert them into smoking pipes and loads of them are sent home to their friends."[10]

The physical destruction of plants caused by hundreds of thousands of men trampling across the landscape for four years with their associated hordes of livestock and implements of war is easy to imagine but hard to quantify. Participants often wrote of botanical devastation associated with an actual battle.

Assistant Surgeon Dr. Daniel M. Holt, 121st New York, at Spotsylvania Courthouse, Virginia, on May 16, 1864: "Trees are perfectly riddled with bullets. Perhaps you will think it extravagant when I tell you that I have seen trees at least sixteen inches in thickness, and oak ones at that, cut completely off by these leaden messengers. In the advanced rifle pits behind which we fight, every tree is like a brush broom!"[11]

In an arena of war, the mobility of many animal species is a benefit. The ability of some animals to run, swim, or fly away from danger is a readily apparent advantage. Wild animals were not exposed to intense, large-scale exploitation during the Civil War, as were plants in the demand for wood, wood products, and pharmaceuticals. Some animals, such as whales, furbearers, fish, and waterfowl, were pursued commercially but probably on a smaller scale than before or after the conflict. Individuals, citizens and combatants alike, eagerly sought fish and seafood opportunistically for food. Other species, such as squirrels, deer, rabbits, raccoons, opossums, turtles, waterfowl, gamebirds, and even songbirds, were eaten occasionally.

Private J. H. Puckett, 7th Arkansas Infantry, in a letter to his wife from near Shelbyville, Tennessee, on Feb. 10, 1863: "When night came, I could hardly believe my eyes. As far as I could see the heavens were blackened with these little Robins coming in to the Cedar brakes to roost. . . . We lit our torches and went thrashing through the bushes from one tree to another . . . [we] caught . . . in all about 50 . . . I felt amply repaid when we set down to a big chicken stew made of Robins and they were the fattiest things I ever saw of the feathered tribe."[12]

Within the area impacted by the Civil War, there are no documented cases of historical extinctions of plants or animals before or during the conflict. Significant, lasting disruption, if at all, likely occurred only in very sensitive and specialized ecosystems like caves. Intensive guano mining activities probably changed cave environments and resulted in the loss of bat colonies, perhaps for many years. If a stream flowed from a cave, populations of endemic fish, amphibians, and invertebrates may have succumbed. Otherwise, the Civil War caused no long-term impacts to flora and fauna of the region. Disturbance to most ecosystems in the form of fire, flood, and wind is a natural process. Plant and animal populations exhibit remarkable resilience in their abilities to recover from disturbance. In some cases, war-related disturbance set back successional stages. A battle or winter-quarters encampment may have destroyed a mature hardwood forest with fox squirrels and red-shouldered hawks *(Buteo lineatus)*, but the resulting thicket of vigorous saplings favored healthy populations of gray squirrels and red-tailed hawks *(Buteo jamaicensis)*. From an ecological perspective, the change was less dramatic than that caused by an intense hurricane.

Other environmental impacts of the war on wild flora and fauna continued for at least two decades after the conflict, although none were substantial on a large scale with one possible exception. One historian notes that the drastically skewed demographics caused by the war in some areas of the South may have had environmental consequences. In Mississippi, for example, almost half of the white males between fifteen and forty-five were dead or missing at the end of the war.[13] It is easy to imagine that a sudden dearth of labor in an area dominated by small farms would result in many fields being abandoned and reverting to forest through natural regeneration. The war consumed most of the horses and mules that would have otherwise been available as draft animals for southern farmers for a number of years. Their loss also contributed to the abandonment of farmland. To a lesser degree, some farmland was likely deserted because of the impracticality of removing breastworks, trenches, tunnels, and other troublesome debris of war. In the same vein, free-range cattle and hogs, common in the South before the war, are disruptive to natural ecosystems, often eliminating some native plants from impacted areas. The dramatic reduction of livestock during the conflict would allow recovery of some habitats and delay the eventual degradation of others. Cane, once a unique component of some southern ecosystems and very susceptible to overgrazing, is an example of one species that may have temporarily benefited from a war-induced reduction of livestock. Similar impacts in the North were avoided because of a much greater population, burgeoning immigration, and the fact that, except for Gettysburg, the area was spared major campaigns.

An indirect impact of the Civil War on flora and fauna that is ripe for speculation involves the conflict's influence on prominent naturalists of the day and the degree to which their work to enhance the knowledge and appreciation of nature was affected. For example, famed naturalist Louis Agassiz saw his work to establish Harvard's Museum of Comparative Zoology delayed by the war.[14] South Carolina clergyman, naturalist, and Audubon collaborator John Bachman was severely beaten by Union soldiers resulting in a paralyzed arm.[15] Spencer F. Baird, assistant secretary (and later secretary) of the Smithsonian Institution, worked to see that the national museum was not expropriated for military use during the war.[16] Preeminent Harvard botanist Asa Gray was a devout antislavery activist. In 1862 he wrote to Charles Darwin, "I do not do so much scientific work as before the war, but still I keep pottering away."[17] Dr. Jared Kirtland was a physician and the Midwest's most famous naturalist of the time. He spent the war examining the fitness of recruits for military duty and donated his pay to the Soldier's Aid Society.[18] John Muir, wilderness preservationist and founder of the Sierra Club, may have avoided the Union draft by going to Canada.[19] These men and others made significant contributions to understanding the wild flora and fauna of America before and after the Civil War. Absent the strife, their achievements would have surely been different.

Impacts of Flora and Fauna on the Civil War

No less than today, the economy of the country during the Civil War was dependent on the exploitation of natural resources. Flora in the form of trees and their products was a prime target. Simply stated, the war could not have occurred as it did without trees. The scope of a war without tree products can be imagined if one considers a major conflict of the same era occurring in the North African deserts or treeless steppes of Asia. From use as rifle stocks to shade (which was mentioned on a dozen occasions in the anecdotes quoted in this book), wild plants defined the Civil War.

Many sources refer to the impacts that plants in general had on an event during the war. One modern historian states, "Perhaps more so than any other battlefields, vegetation shaped the nature of the soldiers' experiences at both Chancellorsville and the Wilderness (as well as the decision making of Hooker, Grant, Meade, and Lee). The vegetation there was commented on almost universally in hundreds of soldiers' letters and memoirs."[20] Another writes, "The terrain and vegetation on the [Stones River] battlefield had a significant impact on the outcome of the battle. The cedar and hardwood forests on the battlefield significantly hindered troop movements and provided cover for some units during the fighting."[21]

Examples of cases where a single, identifiable species of plant impacted an event in the war are much less common. Osage orange (*Maclura pomifera*) is a small, thorn-laden tree once native to the Red River Valley. Local Osage Indians valued the strong, elastic wood for bows and extended its range by trading with eastern tribes. White settlers quickly learned to use the bristling trees to create impenetrable living fences before the invention of barbed wire. On June 7, 1863, Confederate Major General J. G. Walker attacked a Union force at Milliken's Bend in Madison Parish, Louisiana, in hopes of relieving pressure on the besieged fortress of Vicksburg. His attack was thwarted, in part, because of a dense Osage orange hedge around part of the village. Yankees massed behind the hedge and fired through the openings. General Walker declared, "Upon reaching the hedges it was utterly impracticable to pass them except through the few openings left for convenience by the planter. In doing this, the order of battle was necessarily broken.... Owing to these frequent interruptions in the advance in the order of battle ... the ensemble of the movement upon the enemy's position was thus necessarily lost...."[22] Consequently, a battle was lost because of Osage orange.

The Civil War pharmacopoeia comprised many plants that had no beneficial effects on diseased or injured patients, and some were harmful. However, one more than any other species saved thousands of lives. Cinchona, a small tree native to the Andes and the source of quinine, may have been the single most important plant in the war. Almost a million cases of malaria were recorded just among Union troops, and entire units were incapacitated at times. Nothing relieved malarial fevers and returned men to duty like quinine. The Union, in possessing the only two quinine factories in the country and implementing an effective blockade of the South, maintained a decided advantage in the battle for troop fitness.[23]

Wild plants, notably berries and nuts, were often used as food in the Civil War. Soldiers' pervasive fondness for blackberries resulted in seasonal prophylactic doses of Vitamin C that kept scurvy at bay. The impacts of wild flora as food were otherwise minor and local.

Animals beyond the scope of this book, including domesticated cattle, horses, mules, pigs, and fowl, fed soldiers and citizens and moved and equipped armies. Wild fauna impacted the Civil War by dealing out misery, sickness, and death to people in the form of mosquitoes, body lice, and ticks with the diseases they transmitted. In some instances wild animals or their products (e.g., honey, turtle eggs) provided a food source, a significant benefit only on a small scale. The group of fauna that rivals pestilent insects as having had the greatest impact on the Civil War is cave-dwelling bats. For the Confederacy, guano mined in bat caves was the major source of potassium nitrate—the prime component of gunpowder. Without bats and the subsequent guano, the Confederacy would have run short of gunpowder long before it did, and the ability to sustain the war effort was directly dependent on the availability of munitions.

Intangible impacts of wild flora and fauna on the Civil War include the aesthetic balm provided to people in a time when anxiety levels were high and emotions taut.

Refugee Kate Stone, near Monroe, Louisiana, on April 21, 1863: "The profusion of catalpa trees, all in full bloom, lining the streets of Monroe was indescribably fair in the early morning light. The deep green leaves seemed heaped with pyramids of snow. We never thought the catalpa could be so pretty."[24]

Private Isaac Jackson, 17th Ohio Battery, during the siege of Vicksburg, Mississippi, on June 28, 1863: "Among other things, I love to sit and listen to the mocking bird sing. They are plenty here and the pretties singers I ever heard, and they never get tired. There used to be one came where our gun used to [be] planted and sit on an old, dead tree and sing. I could lay in my tent and see him. It seemed as if it never would get tired of singing."[25]

1865 Onward

The aftermath of the Civil War left a country wallowing in reconstruction and soon to begin a broad-based assault on wild flora and fauna of North America. Two factors fueled the destruction: new technology and the rush to rebuild the South. Both resulted in the near total elimination of virgin forests in the eastern United States. Wildlife was not spared. Within two generations of the war, regional and global extinctions of such species as eastern elk (*Cervus canadensis*), red wolves (*Canis rufus*), ivory-billed woodpeckers, whooping cranes (*Grus americana*), Carolina parakeets, and passenger pigeons occurred. The saga of American bison, plume-bearing wading birds, East Coast fisheries, and waterfowl is well documented. Wildlife populations plunged because of loss of habitat and direct overexploitation such as market hunting that accelerated with the development of railroads and refrigeration. The biblical mandate to subdue the earth was pursued in earnest, and few protested.

The tide began to turn after misuse of natural resources of the eastern

United States peaked, and large-scale clearing of western forests along with the slaughter of millions of bison began. Enough influential people valued a frontier of some sort that the preservationist movement was born to set aside and protect at least a small part of it. The era of conservation followed with intensive efforts to manage and enhance the remaining populations of plants and animals, especially those with monetary or sporting value. The present era of conservation biology attempts to address flora and fauna on a broad landscape scale encompassing all ecosystems on the planet. In the history of humans, flora, and fauna, the challenges are unprecedented. Issues are complicated with factions as partisan as those in the American Civil War. Only a successful outcome will prevent resource-based wars of the future and their unimaginable environmental consequences.

General Robert E. Lee, CSA, in a letter to his wife from Pocahontas County, present-day West Virginia, on Aug. 4, 1861: "I enjoyed the mountains, as I rode along. The views are magnificent—the valleys so beautiful, the scenery so peaceful. What a glorious world Almighty God has given us. How thankless and ungrateful we are, and how we labour to mar his gifts."[26]

NOTES

Introduction

1. Kirby, "The American Civil War."
2. Higginson, *Army Life in a Black Regiment*, 5.
3. Smith and Baker, eds., *Burning Rails as We Pleased*, 69.
4. Speer, *Voices from Cemetery Hill*, 96. This is a reference to American shad (*Alosa sapidissima*).
5. Hammer, *Frederic Augustus James's Civil War Diary*, 42–43. Broomsedge bluestem (*Andropogon virginicus*) and closely related species were used to make brooms.
6. Franklin, *The Diary of James T. Ayers*, 72.
7. Fehrenbacher, *Abraham Lincoln: Speeches and Writings, 1859–1865*, 346. Elderberries (*Sambucus* spp.) have pithy stems that are easily hollowed and fashioned into popgun and water-squirt toys.

The Civil War Setting

1. Williams, *Americans and Their Forests*, 3.
2. Ibid., 119–20.
3. Cowdrey, *This Land, This South*, 93.
4. Earley, *Looking for Longleaf*, 83.
5. Williams, *Americans and Their Forests*, 158.
6. Whitney, *From Coastal Wilderness to Fruited Plain*, 147.
7. Peabody, *Aesthetic Papers*, 231.
8. National Park Service, *Historical Land Cover/Use Classification of Pea Ridge National Military Park*.
9. Goodwin, "Big Game Animals in the Northeastern United States."
10. Gabrielson, *Wildlife Management*, 56–62.

11. U.S. Bureau of the Census.
12. U.S. Census Data, "Population of the United States—1860."

PART I: FLORA

INTRODUCTION

1. Adams, *Doctors in Blue*, 39.
2. Garrison and Garrison, *The Encyclopedia of Civil War Usage*, 249.
3. Moore, "Standard Supply Table of the Indigenous Remedies for Field Service and the Sick in General Hospitals," 1–7.
4. Cunningham, *Doctors in Gray*, 147–48.
5. Moseley, *The Stilwell Letters*, 165. The young, tender greens of American pokeweed (*Phytolacca americana*) were often eaten.
6. Silver, *A Life for the Confederacy*, 76. Black haw is the common name for the small tree or shrub known scientifically as *Viburnum prunifolium*.
7. Runge, *Four Years in the Confederate Artillery*, 66.
8. Anderson, *Brokenburn*, 318. The candles were likely made from prickly pear (*Opuntia* sp.), a common cactus of east Texas.
9. Grimsley and Miller, *The Union Must Stand*, 79.
10. Westervelt and Palladino, *Diary of a Yankee Engineer*, 219.
11. Furry, *The Preacher's Tale*, 64.
12. Sauers, *The Civil War Journal of Colonel William J. Bolton*, 199.
13. Burlingame and Ettlinger, *Inside Lincoln's White House*, 177.

ASH

1. Allen, *Trees, Shrubs, and Woody Vines of Louisiana*, 102.
2. Harrar and Harrar, *Guide to Southern Trees*, 623–24; Sargent, *Manual of the Trees of North America*, 842; Thomas, *100 Woody Plants of North Louisiana*, 164.
3. Allen, *Trees, Shrubs, and Woody Vines of Louisiana*, 103–104.
4. Porcher, *Resources of the Southern Fields and Forests, Medical, Economical, and Agricultural*, 494. Hereafter cited as "Porcher."
5. Jackson, *Some of the Boys*, 22.
6. Taylor, *Reluctant Rebel*, 63.
7. Allen, *Trees, Shrubs, and Woody Vines of Louisiana*, 66.
8. "Wadley Diary," 209.

BALDCYPRESS

1. Irion et al., *Remote Sensing Investigations of Civil War Era Shipwrecks in the Vicinity of Fort St. Philip, Plaquemines Parish, Louisiana*, 1.
2. Moore, *Andrew Brown and Cypress Lumbering in the Old Southwest*, 153, 156.
3. Heidler, Heidler, and Coles, *Encyclopedia of the American Civil War*, 1412.

4. Porcher, 509.

5. "Wadley Diary," 89–90.

6. Angle, *Three Years in the Army of the Cumberland,* 328.

7. Grimsley and Miller, *The Union Must Stand,* 82.

8. "Wadley Diary," Vol. 2, 81.

9. Clark, *Downing's Civil War Diary,* 137.

10. Sears, *For Country, Cause & Leader,* 210.

11. Larimer, *Love and Valor,* 70.

12. Grimsley and Miller, *The Union Must Stand,* 219.

13. Texas Confederate Military Organizations.

BEECH

1. Porcher, 236.

2. Allen, *Trees, Shrubs, and Woody Vines of Louisiana,* 98; Porcher, 235.

3. Garrison and Garrison, *The Encyclopedia of Civil War Usage,* 217; Harrar and Harrar, *Guide to Southern Trees,* 166; Sargent, *Manual of the Trees of North America,* 229; Porcher, 235.

4. Porcher, 237.

5. Andersen, *The Civil War Diary of Allen Morgan Geer, 20th Regiment Illinois Volunteers,* 138–39.

6. Coe, *Mine Eyes Have Seen the Glory,* 28.

7. Davis, *Diary of a Confederate Soldier,* 73.

8. Cutrer and Parrish, *Brothers in Gray,* 22–23.

9. Sears, *On Campaign with the Army of the Potomac,* 67.

10. Larimer, *Love and Valor,* 63.

11. Wiley, *This Infernal War,* 152.

BLACKBERRY

1. Rankin, *Diary of a Christian Soldier*

2. Sherman, *Memoirs of General William T. Sherman,* 392.

3. Porcher, 141–44.

4. Quaife, *From the Cannon's Mouth,* 242.

5. Swedberg, *Three Years with the 92nd Illinois,* 99, 218–19.

6. Roth, *Well Mary,* 80.

7. Rankin, *Diary of a Christian Soldier,* 97.

8. Sears, *For Country, Cause & Leader,* 36.

9. Cumming, *Kate,* 110–11.

10. Davis, *Diary of a Confederate Soldier,* 46.

11. Greiner et al., *A Surgeon's Civil War,* 125.

12. Westervelt and Palladino, *Diary of a Yankee Engineer,* 151.

13. Winther, *With Sherman to the Sea,* 122.

14. Jones, *The Civil War Memoirs of Captain William J. Seymour,* 81–82.

15. Dawson, *A Confederate Girl's Diary,* 52.

16. Grimsley and Miller, *The Union Must Stand,* 48.

17. Snell, *Myra Inman,* 105.

18. Bauer, *Soldiering,* 130–31.

19. Larimer, *Love and Valor,* 204.

20. McDonald, *Make Me a Map of the Valley,* 63.

21. Clark, *Downing's Civil War Diary,* 55–56.

22. Fehrenbacher, *Abraham Lincoln: Speeches and Writings, 1859–1865,* 315.

23. Garrison and Garrison, *The Encyclopedia of Civil War Usage,* 26.

CANE

1. Roosevelt, "In the Louisiana Canebrakes."

2. Robin, *Voyage to Louisiana, 1803–1805,* 128.

3. Anderson and Anderson, *The Generals,* 449.

4. Sauers, *The Civil War Journal of Colonel William J. Bolton,* 124.

5. Northern, *All Right Let Them Come,* 61.

6. Sears, *For Country, Cause & Leader,* 332.

7. Grimsley and Miller, *The Union Must Stand,* 105.

8. Maynard, "Vicksburg Diary," 51.

9. Andersen, *The Civil War Diary of Allen Morgan Geer,* 137.

10. Cutrer and Parrish, *Brothers in Gray,* 151.

11. U.S. War Department, *The War of the Rebellion: A Compilation of the Official Records of the Union and Confederate Armies,* Vol. 24, 184. Hereafter cited as OR. Sap rollers were dense cylindrical baskets rolled ahead of men working on a trench to provide cover from small-arms fire.

12. Sanders, "Diary," 208.

13. Johnson and Buel, *Battles and Leaders of the Civil War,* Vol. 3, 492.

14. Greene, *The Civil War Diary of Lieutenant Robert Molford Addison,* 60.

15. Craig, "Civil War Letters," letter of Sept. 8, 1863.

16. Anderson, *Brokenburn,* 166.

17. OR, Vol. 34, 689.

18. Byrne, *The View from Headquarters,* 39.

19. Robertson Jr., *Soldier of Southwestern Virginia,* 65.

20. OR, Vol. 34, 874.

CHESTNUT AND CHINQUAPIN

1. Southgate, "Historical Ecology of American Chestnut (*Castanea dentata*)," 1.

2. Harrar and Harrar, *Guide to Southern Trees,* 166.

3. Sargent, *Manual of the Trees of North America,* 232.

4. Harrar and Harrar, *Guide to Southern Trees,* 170.

5. Porcher, 238.

6. McDonald, *Make Me a Map of the Valley,* 106.

7. OR, Vol. 34, 358.

8. Moseley, *The Stilwell Letters*, 232.

9. Pelka, *The Civil War Letters of Colonel Charles F. Johnson, Invalid Corps*, 51.

10. Andersen, *The Civil War Diary of Allen Morgan Geer*, 59.

11. Winther, *With Sherman to the Sea*, 131.

12. Snell, *Myra Inman*, 169.

13. Angle, *Three Years in the Army of the Cumberland*, 270.

14. Greiner et al., *A Surgeon's Civil War*, 140.

15. Brown, *One of Cleburne's Command*, 111.

16. Larimer, *Love and Valor*, 280.

17. Rosenblatt, *Hard Marching Every Day*, 8.

18. Sears, *For Country, Cause & Leader*, 222.

19. Child, *Letters from a Civil War Surgeon*, 135.

20. Taylor, *Reluctant Rebel*, 240.

21. "Wadley Diary," 38.

22. Wiley, *This Infernal War*, 142.

23. Silver, *A Life for the Confederacy*, 49.

24. Peel, Civil War Diary, Sept. 22, 1861.

25. Brown, *One of Cleburne's Command, CSA*, 82.

CINCHONA

1. Cinchona. Tropical Plant Database, 482.

2. Bollet, *Civil War Medicine*, 289.

3. Flannery, *Civil War Pharmacy*, 118, 131.

4. Adams, *Doctors in Blue*, 140.

5. Flannery, *Civil War Pharmacy*, 164.

6. Jacobs, "Some of the Drug Conditions During the War Between the States, 1861–5."

7. Gaines, *Encyclopedia of Civil War Shipwrecks*, 125.

8. Nash, "Some Reminiscences of a Confederate Surgeon," 133.

9. Porcher, 39, 59, 238, 334, 412. *Boneset* is the colloquial name for several herbs in the *Eupatorium* genus.

10. Massey, *Ersatz in the Confederacy*, 120.

11. Adams, *Doctors in Blue*, 219.

12. Billings, *Hardtack and Coffee*, 175–76.

13. Quaife, *From the Cannon's Mouth*, 336.

14. Silver, *A Life for the Confederacy*, 46–47.

15. Anderson, *Brokenburn*, 64.

16. Pelka, *The Civil War Letters of Colonel Charles F. Johnson, Invalid Corps*, 118.

17. Holland, *Keep All My Letters*, 98.

18. Cash and Howorth, *My Dear Nellie*, 99.

19. Alspaugh, Letters.

20. Miers, *A Rebel War Clerk's Diary*, 284.

21. Radigan, *Desolating This Fair Country*, 74.

22. Crawford, *William Howard Russell's Civil War*, 112.

23. "Wadley Diary," Vol. 2, 54.

24. Grimsley and Miller, *The Union Must Stand*, 14.

25. Cutrer and Parrish, *Brothers in Gray*, 47–48.

26. Sears, *On Campaign with the Army of the Potomac*, 12.

27. Durham, *The Blues in Gray*, 44–45.

28. Child, *Letters from a Civil War Surgeon*, 57.

29. Holcomb, *Southern Sons, Northern Soldiers*, 24. Ipecac syrup was made from the Brazilian plant ipecacuanha (*Psychotria ipecacuanha*), and opium is derived from opium poppies (*Papaver somniferum*).

30. Basile, *The Civil War Diary of Amos E. Stearns*, 32.

COTTONWOOD

1. Miller et al., *The Struggle for the Life of the Republic*, 65.

2. Hunter and Hunter, *Steamboats on the Western Rivers*, 265.

3. Bragg et al., *Never for Want of Powder*, 77.

4. Corbusier and Wooster, *Soldier, Surgeon, Scholar*, 53.

5. Jackson, *Some of the Boys*, 217.

6. Anderson, *Memories*, 338–39.

DOGWOOD

1. Porcher, 61–62.

2. Thomas, *100 Woody Plants of North Louisiana*, 58.

3. Sargent, *Manual of the Trees of North America*, 786; Porcher, 62; Thomas, *100 Woody Plants of North Louisiana*, 58.

4. "Wadley Diary," Vol. 2, 190–91.

5. Angle, *Three Years in the Army of the Cumberland*, 198.

6. Crawford, *William Howard Russell's Civil War*, 38.

7. Holland, *Keep All My Letters*, 85.

8. Anderson, *Brokenburn*, 226.

ELM

1. Sargent, *Manual of the Trees of North America*, 310, 314; Porcher, 310–11; Harrar and Harrar, *Guide to Southern Trees*, 234.

2. Bearss, *A Louisiana Confederate*, 32–33.

3. Lincecum et al., *Gideon Lincecum's Sword*, 221.

GRAPE

1. Martin et al., *American Wildlife & Plants*, 344.

2. Porcher, 229–30.

3. Allen, "The Paper Money of the Confederate States," 7.

4. Winther, *With Sherman to the Sea*, 140.

5. Anderson, *Brokenburn,* 53.

6. Davis, *Diary of a Confederate Soldier,* 84. Muscadines ripen in early autumn; this was likely another grape species.

7. Menge and Shimrak, *The Civil War Notebook of Daniel Chisholm,* 12.

8. Snell, *Myra Inman,* 168.

9. Lowe, *A Texas Cavalry Officer's Civil War,* 4.

10. Crawford, *William Howard Russell's Civil War,* 55.

11. Stauffer, *Civil War Diary,* n.p.

12. Greiner et al., *A Surgeon's Civil War,* 29.

13. Coe, *Mine Eyes Have Seen the Glory,* 163.

14. Pelka, *The Civil War Letters of Colonel Charles F. Johnson, Invalid Corps,* 82.

15. McDonald, *Make Me a Map of the Valley,* 91.

16. Robertson Jr., *Soldier of Southwestern Virginia,* 83.

17. Garrison and Garrison, *The Encyclopedia of Civil War Usage,* 98.

18. Gaines, *Encyclopedia of Civil War Shipwrecks,* 128.

HERBS

1. Porcher, 440; Foster and Tyler, *Tyler's Honest Herbal,* 293.

2. "Wadley Diary," Vol. 2, 83–84. Laudanum is an opium derivative mixed with ethyl alcohol.

3. Cumberworth and Biles, *An Enduring Love,* 130. Oil of wintergreen (methyl salicylate) is produced in several groups of plants including those in genera *Pyrola, Gaultheria,* and *Betula.*

4. Sutherland, *A Very Violent Rebel,* 143. Catnip is the common name of several species of *Nepeta.*

5. Skipper and Taylor, *A Handful of Providence,* 150. Horsemint may be a reference to European *Mentha longifolia* or any species of the native genus *Monarda.*

6. Porcher, 429–31; Foster and Tyler, *Tyler's Honest Herbal,* 137.

7. Sears, *On Campaign with the Army of the Potomac,* 63.

8. Cumberworth and Biles, *An Enduring Love,* 166.

9. Foster and Tyler, *Tyler's Honest Herbal,* 319.

10. Stauffer, *Civil War Diary.* Pages unnumbered.

11. Porcher, 20; Foster and Tyler, *Tyler's Honest Herbal,* 51.

12. Patch, *This from George,* 102.

13. Porcher, 86; Foster and Tyler, *Tyler's Honest Herbal,* 353.

14. Lowe, *A Texas Cavalry Officer's Civil War,* 292–93.

15. Cullina, *The New England Wild Flower Society Guide to Growing and Propagating Wildflowers of the United States and Canada,* 154.

16. Snell, *Myra Inman,* 194.

17. Porcher, 401; Foster and Tyler, *Tyler's Honest Herbal,* 249.

18. Cutrer, *Oh, What a Loansome Time I Had,* 107.

19. Porcher, 511–16; Gaines, *Encyclopedia of Civil War Shipwrecks,* 47, 122.

20. Cumming, *Kate,* 28.

21. Brumgardt, *Civil War Nurse,* 105.

22. Howland and Bacon, *My Heart Toward Home,* 158.

23. Porcher, 532; Foster and Tyler, *Tyler's Honest Herbal,* 171.

24. OR, Vol. 25/2, 687.

25. Hammer, *Frederic Augustus James's Civil War Diary,* 67.

26. Peel, Civil War Diary, May 8, 1863.

27. Moseley, *The Stilwell Letters,* 145.

HUCKLEBERRY

1. Porcher, 384.

2. Husby and Wittenberg, *Under Custer's Command,* 46.

3. Moseley, *The Stilwell Letters,* 18–19.

4. Stauffer, *Civil War Diary,* n.p.

5. Menge and Shimrak, *The Civil War Notebook of Daniel Chisholm,* 133.

6. Grimsley and Miller, *The Union Must Stand,* 46.

7. Rosenblatt, *Hard Marching Every Day,* 30.

JUNIPER

1. Sargent, *Manual of Trees of North America,* 88; Anderson and Anderson, *The Generals,* 338.

2. Porcher, 510.

3. Davis, *Diary of a Confederate Soldier,* 69.

4. Reddick, *Seventy-Seven Years in Dixie,* 31.

5. Grimsley and Miller, *The Union Must Stand,* 132.

6. Smith and Baker, *Burning Rails as We Pleased,* 24.

7. Dennis, *Kemper County Rebel,* 56.

8. Angle, *Three Years in the Army of the Cumberland,* 55–56.

9. Rosenblatt, *Hard Marching Every Day,* 278.

10. Smith and Baker, *Burning Rails as We Pleased,* 42–43.

11. Sears, *For Country, Cause and Leader,* 34–35.

12. Cumming, *Kate,* 272.

13. Brown, *One of Cleburne's Command,* 153.

14. Miers, *A Rebel War Clerk's Diary,* 320.

LOCUST

1. Thomas, *100 Woody Plants of North Louisiana,* 76.

2. Porcher, 191–92.

3. Edlin, *The Tree Key,* 156.

4. Thomas, *100 Woody Plants of North Louisiana,* 76.

5. Sargent, *Manual of the Trees of North America,* 609; Porcher, 195; Cassidy and Hall, *Dictionary of American Regional English,* Vol. 1, 750.

6. Smith and Cooper, *A Union Woman in Civil War Kentucky,* 161.

7. Snell, *Myra Inman,* 182.

8. Reddick, *Seventy-Seven Years in Dixie,* 29.

9. Larimer, *Love and Valor,* 142.

10. Gettysburg National Military Park, News Release, 18 Aug. 2008.

MAGNOLIA

1. Porcher, 36–41.

2. Thomas, *100 Woody Plants of North Louisiana,* 138, 142.

3. Franklin, *The Diary of James T. Ayers,* 48.

4. Angle, *Three Years in the Army of the Cumberland,* 380, 383.

5. Howland and Bacon, *My Heart Toward Home,* 232.

6. Winther, *With Sherman to the Sea,* 145.

7. Kiper, *Dear Catharine, Dear Taylor,* 342.

8. "Magnolia Hall."

MAPLE

1. "Rope-tension drums," National Music Museum.

2. "Maple sugaring history, Naper Settlement."

3. Sargent, *Manual of the Trees of North America,* 689.

4. Porcher, 79.

5. Lee, *Recollections and Letters of General Robert E. Lee,* 42.

6. Sears, *On Campaign with the Army of the Potoma,* 147.

7. Jackson, *Some of the Boys,* 21.

8. "*Maple Leaf* Shipwreck," National Park Service.

MISTLETOE

1. Porcher, 63.

2. Howland and Bacon, *My Heart Toward Home,* 128.

3. Anderson, *Brokenburn,* 164. "Casino berries" may be a reference to deciduous holly (*Ilex decidua*) or dahoon holly (*I. cassine*), both of which have bright red berries in the winter.

4. Holcomb, *Southern Sons, Northern Soldier,* 55.

5. Follett, letter of April 25, 1862.

6. Looby, *The Complete Civil War Journal and Selected Letters of Thomas Wentworth Higginson,* 104.

7. *Official Records of the Union and Confederate Navies in the War of the Rebellion,* Ser. II, Vol. 1, 146.

MULBERRY

1. U.S. Forest Service, "Gypsy Moth in North America."

2. U.S. Forest Service, "White Mulberry."

3. Porcher, 307.

4. Ibid., 305.

5. Angle, *Three Years in the Army of the Cumberland,* 217.

6. Anderson, *Brokenburn,* 19.

7. Greiner et al., *A Surgeon's Civil War*, 206.

8. Smith and Baker, *Burning Rails as We Pleased*, 56.

9. Burlingame and Ettlinger, *Inside Lincoln's White House*, 59.

10. "Civil War Journal from the 4th Iowa Cavalry," May 19, 1863.

OAKS

1. Thomas, *100 Woody Plants of North Louisiana*, 80.

2. Porcher, 257, 259.

3. Jacobs, "Some of the Drug Conditions During the War Between the States, 1861–5."

4. Lincecum et al., *Gideon Lincecum's Sword*, 222.

5. Porcher, 239, 262; Thomas, *100 Woody Plants of North Louisiana*, 82, 92, 94; Sargent, *Manual of the Trees of North America*, 238, 301.

6. Jacobs, "Some of the Drug Conditions During the War Between the States, 1861–5."

7. Porcher, 239, 257.

8. Still, *Confederate Shipbuilding*, 58.

9. Konstam, *Mississippi River Gunboats of the American Civil War 1861–1865*, 6, 15.

10. Porter, *The Naval History of the Civil War*, 364.

11. Cumming, *Kate*, 129.

12. Angle, *Three Years in the Army of the Cumberland*, 220.

13. Cash and Howorth, *My Dear Nellie*, 117–18.

14. Wiley, *This Infernal War*, 272.

15. Rosenblatt, *Hard Marching Every Day*, 164.

16. Cutrer and Parrish, *Brothers in Gray*, 62.

17. Miers, *A Rebel War Clerk's Diary*, 255.

18. Lowe, *A Texas Cavalry Officer's Civil War*, 136–37.

19. Sears, *On Campaign with the Army of the Potomac*, 154–55.

20. Clark, *Downing's Civil War Diary*, 83.

21. Runge, *Four Years in the Confederate Artillery*, 71.

22. Byrne, *The View from Headquarters*, 19.

23. Lowe, *A Texas Cavalry Officer's Civil War*, 34. "Over cap" acorns in this area of Oklahoma is likely a reference to bur oak (*Quercus macrocarpa*).

24. *Confederate Receipt Book*, 16.

25. Sears, *On Campaign with the Army of the Potomac*, 291.

26. Lane, *Dear Bet*, 70.

27. Cumming, *Kate*, 280.

28. Wiley, *This Infernal War*, 359.

29. Sears, *For Country, Cause & Leader*, 279.

30. Child, *Letters from a Civil War Surgeon*, 247.

31. Kiper, *Dear Catharine, Dear Taylor*, 149.

32. Sears, *For Country, Cause & Leader*, 209.

33. Franklin, *The Diary of James T. Ayers*, 80.

34. Byrne, *The View from Headquarters,* 214.

35. Anderson, *Brokenburn,* 222. "Blackjack" is a reference to blackjack oak (*Quercus mari-landica*).

36. Greiner et al., *A Surgeon's Civil War,* 232.

37. Grimsley and Miller, *The Union Must Stand,* 129.

38. Menge and Shimrak, *The Civil War Notebook of Daniel Chisholm,* 157.

39. Chesnut, *A Diary from Dixie,* 409.

40. Rankin, *Diary of a Christian Soldier,* 171.

41. Lowe, *A Texas Cavalry Officer's Civil War,* 18.

42. Brown, *One of Cleburne's Command,* 6–7.

43. Ibid., 74.

44. Elder, *Love Amid the Turmoil,* 247.

45. Child, *Letters from a Civil War Surgeon,* 25.

46. Holcomb, *Southern Sons, Northern Soldiers,* 7.

47. Clark, *Downing's Civil War Diary,* 39.

PALMETTO

1. Porcher, 526.

2. Ibid., 525–26.

3. Cotton, *Yankee Bullets, Rebel Rations,* 41.

4. "Wadley Diary," Vol. 1, 99.

5. Cumming, *Kate,* 249.

6. Anderson, *Brokenburn,* 48.

7. Grimsley and Miller, *The Union Must Stand,* 86.

8. Looby, *The Complete Civil War Journal and Selected Letters of Thomas Wentworth Higginson,* 46.

9. "Wadley Diary," Vol. 2, 70.

10. Holcomb, *Southern Sons, Northern Soldiers,* 55.

11. OR, Vol. 26, 224.

12. OR, Vol. 35, 113.

13. Durham, *The Blues in Gray,* 43.

14. OR, Vol. 48, 58.

15. OR, Vol. 35, 41.

16. Westervelt and Palladino, *Diary of a Yankee Engineer,* 2–3.

17. Porcher, 526.

18. Westervelt and Palladino, *Diary of a Yankee Engineer,* 6–7.

PERSIMMON

1. Sargent, *Manual of the Trees of North America,* 822; Hague, *A Blockaded Family,* 70.

2. Porcher, 386.

3. McGuire, "Progress of Medicine in the South."

4. Garrison and Garrison, *The Encyclopedia of Civil War Usage*, 190.

5. Sears, *For Country, Cause & Leader*, 94.

6. Angle, *Three Years in the Army of the Cumberland*, 326.

7. Miers, *A Rebel War Clerk's Diary*, 450.

8. Davis, *Diary of a Confederate Soldier*, 93.

9. Larimer, *Love and Valor*, 238.

10. OR, Vol. 21, 831.

11. Andersen, *The Civil War Diary of Allen Morgan Geer*, 58.

12. Taylor, *Reluctant Rebel*, 124.

13. Radigan, *Desolating This Fair Country*. Papaws (*Asimina triloba*) have large, sweet, edible fruits.

14. Stauffer, *Civil War Diary*. Pages unnumbered.

15. Porcher, 387.

16. Angle, *Three Years in the Army of the Cumberland*, 273.

PINE

1. Craughwell, *Stealing Lincoln's Body*, 4.

2. Harrar and Harrar, *Guide to Southern Trees*, 51.

3. Garrison and Garrison, *The Encyclopedia of Civil War Usage*, 188.

4. Earley, *Looking for Longleaf*, 27, 87, 91, 96, 99–102; Porcher, 495, 499–501; Sargent, *Manual of the Trees of North America*, 15; Still, *Confederate Shipbuilding*, 58; Gaines, *Encyclopedia of Civil War Shipwrecks*, 38, 62; Hague, *A Blockaded Family*, 42; Porter, *The Naval History of the Civil War*, 763.

5. Smith and Baker, *Burning Rails as We Pleased*, 75.

6. Rosenblatt, *Hard Marching Every Day*, 63.

7. Runge, *Four Years in the Confederate Artillery*, 35.

8. Samito, *Fear Was Not in Him*, 58.

9. Jones, *The Civil War Memoirs of Captain William J. Seymour*, 101–102.

10. Barrett, *Yankee Rebel*, 95.

11. Greiner et al., *A Surgeon's Civil War*, 217.

12. Holland, *Keep All My Letters*, 82.

13. Swedberg, *Three Years with the 92nd Illinois*, 222–23.

14. Dawson, *A Confederate Girl's Diary*, 273.

15. Cumming, *Kate*, 88.

16. Miers, *A Rebel War Clerk's Diary*, 492.

17. Snell, *Myra Inman*, 224.

18. Galbraith, *A Lost Heroine of the Confederacy*, 185.

19. "Wadley Diary," Vol. 1, 56.

20. Jackson, *Some of the Boys*, 246.

21. Hammer, *Frederic Augustus James's Civil War Diary*, 78.

22. Reinhart, *Two Germans in the Civil War*, 142.

23. Bauer, *Soldiering*, 29–30.

24. Eakin and Peoples, *In Defense of My Country*, 145.

25. Runge, *Four Years in the Confederate Artillery*, 68–69.

26. Trimble, *Brothers 'Til Death*, 19.

27. Clark, *Downing's Civil War Diary*, 238.

28. Elliot, *Doctor Quintard, Chaplain C.S.A. and Second Bishop of Tennessee*, 254.

29. Angle, *Three Years in the Army of the Cumberland*, 386.

30. Straubing, *In Hospital and Camp*, 150.

31. Westervelt and Palladino, *Diary of a Yankee Engineer*, 11.

32. Rankin, *Diary of a Christian Soldier*, 171.

33. Husby and Wittenberg, *Under Custer's Command*, 74.

34. Clark, *Downing's Civil War Diary*, 253.

35. Pelka, *The Civil War Letters of Colonel Charles F. Johnson, Invalid Corps*, 52.

36. Runge, *Four Years in the Confederate Artillery*, 131.

37. York, The Civil War Letters of Galutia York, May 16, 1863.

38. Angle, *Three Years in the Army of the Cumberland*, 329–30.

39. Yacovone, *A Voice of Thunder*, 295–96.

40. Davis, *Diary of a Confederate Soldier*, 53–54.

41. Kiper, *Dear Catharine, Dear Taylor*, 398.

42. Smith and Baker, *Burning Rails as We Pleased*, 141–42.

43. Roth, *Well Mary*, 113.

44. Byrne, *The View from Headquarters*, 211.

45. Miers, *A Rebel War Clerk's Diary*, 15. This anecdote may refer to damage caused by black turpentine beetles (*Dendroctonus terebrans*), southern pine beetles (*Dendroctonus frontalis*), or IPS engraver beetles (*Ips* spp.).

46. Rosenblatt, *Hard Marching Every Day*, 166.

47. Bauer, *Soldiering*, 221.

48. Stauffer, *Civil War Diary*. Pages unnumbered.

49. Sears, *For Country, Cause & Leader*, 157.

50. Raab, *With the 3rd Wisconsin Badgers*, 153–54.

51. Larimer, *Love and Valor*, 221.

52. Moseley, *The Stilwell Letters*, 126.

53. Lee, *Recollections and Letters of General Robert E. Lee*, 111.

SASSAFRAS

1. Cowdrey, *This Land, This South*, 29; Porcher, 352; Cunningham, *Doctors in Gray*, 189.

2. Sargent, *Manual of the Trees of North America*, 363; Harrar and Harrar, *Guide to Southern Trees*, 309; Thomas, *100 Woody Plants of North Louisiana*, 134.

3. Jackson, *Some of the Boys*, 90.

4. Cotton, *Yankee Bullets, Rebel Rations*, 55.

5. Holcomb, *Southern Sons, Northern Soldiers*, 8.

6. Anderson, *Brokenburn*, 180.

7. Snell, *Myra Inman*, 301.

8. Larimer, *Love and Valor*, 202.

9. Porcher, 353.

SPANISH MOSS

1. Angle, *Three Years in the Army of the Cumberland*, 329.

2. Sears, *For Country, Cause & Leader*, 332.

3. Franklin, *The Diary of James T. Ayers*, 80.

4. Byrne, *The View from Headquarters*, 213.

5. Bauer, *Soldiering*, 196.

6. Looby, *The Complete Civil War Journal and Selected Letters of Thomas Wentworth Higginson*, 161.

7. Holcomb, *Southern Sons, Northern Soldiers*, 54.

8. Roth, *Well Mary*, 112.

9. Kiper, *Dear Catharine, Dear Taylor*, 341.

10. Lincecum et al., *Gideon Lincecum's Sword*, 125–26.

11. Ibid., 141–42.

12. Larimer, *Love and Valor*, 121.

13. Follett, letter of April 5, 1863.

14. Jackson, *Some of the Boys*, 177.

SUMAC

1. "Mills," Maryland State Archives.

2. *The Leather Manufacturer*. Vol. 9, No. 1.

3. Sargent, *Manual of the Trees of North America*, 663; Thomas, *100 Woody Plants of North Louisiana*, 18, 20; Porcher, 81, 205–207; Jacobs, "Some of the Drug Conditions During the War Between the States, 1861–5."

4. Manarin, *North Carolina Troops, 1861–1865*, Vol. VII, 535.

5. Furry, *The Preacher's Tale*, 6.

SWEETGUM

1. Barrow, Ecology of small insectivorous birds in a bottomland hardwood forest.

2. OR, Ser. 4, Vol. 2, 13.

3. Sears, *For Country, Cause & Leader*, 336.

4. Schwartz, *A Woman Doctor's Civil War*, 117.

5. Fleming, *Civil War and Reconstruction in Alabama*, 238.

6. Hague, *A Blockaded Family*, 105.

7. Snell, *Myra Inman*, 182.

8. Silver, *A Life for the Confederacy*, 91. This account may refer to black gum (*Nyssa sylvatica*), a species with a propensity to form hollows.

9. Stevens, *Centennial History of Missouri*. Vol. 2., 294–95.

10. Peters, *The Underground Railroad in Floyd County, Indiana*, 107.

SYCAMORE

1. Thomas, *100 Woody Plants of North Louisiana,* 168; Allen, *Trees, Shrubs, and Woody Vines of Louisiana,* 184.
2. McGuire, "Progress of Medicine in the South," *Southern Historical Society Papers,* Vol. 17, 11.
3. Winther, *With Sherman to the Sea,* 138.
4. Rosenblatt, *Hard Marching Every Day,* 134.
5. Furry, *The Preacher's Tale,* 63.

WALNUT, HICKORY, AND PECAN

1. Porcher, 320.
2. Ibid., 318–21.
3. Swedberg, *Three Years with the 92nd Illinois,* 137.
4. Miers, *A Rebel War Clerk's Diary,* 205.
5. Cash and Howorth, *My Dear Nellie,* 89. Vicksburg was once known as Walnut Hills.
6. Follett, letter of Feb. 8, 1863.
7. Snell, *Myra Inman,* 178.
8. Ibid., 225.
9. Sears, *For Country, Cause & Leader,* 68. "Hazelnuts" is a reference to American hazelnut (*Corylus americana*), a member of the birch family.
10. Dennis, *Kemper County Rebel,* 49.
11. Child, *Letters from a Civil War Surgeon,* 59.
12. Barrett, *Yankee Rebel,* 9.
13. Porcher, 323–24.
14. Raab, *With the 3rd Wisconsin Badgers,* 234.
15. Wiley, *This Infernal War,* 135–36.
16. Dawson, *A Confederate Girl's Diary,* 233.
17. Dennis, *Kemper County Rebel,* 24. "Scaly boxes" may refer to the nuts of shagbark hickory (*Carya ovata*), also known as scaly bark hickory.
18. Ibid., 30.
19. Cutrer and Parrish, *Brothers in Gray,* 37.
20. Snell, *Myra Inman,* 236.
21. Clark, *Downing's Civil War Diary,* 78.
22. "History of Pecans," National Pecan Sheller's Association.
23. Lincecum et al., *Gideon Lincecum's Sword,* 223.
24. Holcomb, *Southern Sons, Northern Soldiers,* 95.
25. Anderson, *Brokenburn,* 63.
26. Jackson, *Some of the Boys,* 140.

WILLOW

1. Hubbard, *The Basket Willow,* 81.
2. Hague, *A Blockaded Family,* 61 68.

3. Thomas, *100 Woody Plants of North Louisiana*, 188; Porcher, 334–35, 339, 342.
4. Porter, *The Naval History of the Civil War*, 305.
5. Jackson, *Some of the Boys*, 54.
6. Raab, *With the 3rd Wisconsin Badgers*, 249.
7. Jones, *The Civil War Memoirs of Captain William J. Seymour*, 147.

PART II: FAUNA

INTRODUCTION

1. Martin et al., *Biodiversity of the Southeastern United States—Lowland Terrestrial Communities*, 103.
2. Etnier, "Jeopardized Southeastern Freshwater Fishes," 89.
3. Jackson, *Some of the Boys*, 135.
4. Rankin, *Diary of a Christian Soldier*, 90.
5. Stauffer, *Civil War Diary*. Pages unnumbered.
6. Burlingame and Ettlinger, *Inside Lincoln's White House*, 148.
7. Patch, *This from George*, 117.

ALLIGATOR

1. "*Alligator mississippiensis*," Louisiana Alligator Advisory Council.
2. Ibid.
3. Westervelt and Palladino, *Diary of a Yankee Engineer*, 108.
4. Burlingame and Ettlinger, *Inside Lincoln's White House*, 49.
5. Winther, *With Sherman to the Sea*, 161–62.
6. Davis, *Diary of a Confederate Soldier*, 59.
7. Rankin, *Diary of a Christian Soldier*, 97.
8. Earp, *Yellow Flag*, 34.
9. York, The Civil War Letters of Galutia York, Jan. 7, 1863.
10. Ibid., Feb. 1, 1863.
11. Durham, *The Blues in Gray*, 23–24.
12. Eakin and Peoples, *In Defense of My Country*, 123.
13. Rankin, *Diary of a Christian Soldier*, 123.
14. Grimsley and Miller, *The Union Must Stand*, 91.
15. Durham, *The Blues in Gray*, 157.
16. Crawford, *William Howard Russell's Civil War*, 45.
17. Sears, *For Country, Cause & Leader*, 332.
18. Ibid., 334.

BATS

1. Still, *Confederate Shipbuilding*, 49.
2. Bradley, *It Happened in the Civil War*, 2–3.
3. OR, Ser. 4, Vol. 3, 698.

4. Miers, *A Rebel War Clerk's Diary*, 292.
5. Greiner et al., *A Surgeon's Civil War*, 229–30.

BEARS

1. Anderson, *Brokenburn*, 113.
2. Furry, *The Preacher's Tale*, 101.
3. Larimer, *Love and Valor*, 70–71.
4. Jackson, *Some of the Boys*, 217–18.
5. Carmony, *The Civil War in Apacheland*, 86.
6. Garrison and Garrison, *The Encyclopedia of Civil War Usage*, 39.

BIRDS

1. Martin et al., *Biodiversity of the Southeastern United States—Lowland Terrestrial Communities*, 103.
2. Porcher, 390–91.
3. Weaver, *Thank God My Regiment an African One*, 52.
4. Westervelt and Palladino, *Diary of a Yankee Engineer*, 63. Eskimo curlews (*Numenius borealis*) were once one of the most numerous birds in North America but may now be extinct, in part because of market hunting.
5. York, *The Civil War Letters of Galutia York*, Feb. 14, 1863.
6. Sears, *On Campaign with the Army of the Potomac*, 219. Wilson's snipe (*Gallinago delicata*), a type of long-billed, probing sandpiper, is still considered a gamebird today.
7. Anderson, *Brokenburn*, 82.
8. Wiley, *This Infernal War*, 235.
9. Ibid., 433.
10. Lincecum et al., *Gideon Lincecum's Sword*, 292.
11. Chesnut, *A Diary from Dixie*, 366.
12. Peel, Civil War Diary, May 16, 1863.
13. Snell, *Myra Inman*, 291.
14. Miers, *A Rebel War Clerk's Diary*, 351.
15. Cutrer and Parrish, *Brothers in Gray*, 174.
16. Sears, *On Campaign with the Army of the Potomac*, 41.
17. Howland and Bacon, *My Heart Toward Home*, 123.
18. Anderson, *Brokenburn*, 199.
19. Wiley, *This Infernal War*, 422–23.
20. Berlin, *A Confederate Nurse*, 153.
21. Anderson, *Brokenburn*, 70.
22. York, *The Civil War Letters of Galutia York*, Feb. 19, 1863.
23. Durham, *The Blues in Gray*, 58.
24. Sears, *For Country, Cause & Leader*, 342.
25. Pelka, *The Civil War Letters of Colonel Charles F. Johnson, Invalid Corps*, 107.
26. Poe, *The Raving Foe*, 15–16.

27. Civil War Journal from the 4th Iowa Cavalry, July 25, 1863. In this location on this date the "crane" was likely some species of heron or egret, a distinction often disregarded even today.

28. Angle, *Three Years in the Army of the Cumberland,* 217.

29. Chesnut, *A Diary from Dixie,* 170. "Yellow Hammer" is a colloquial name for the northern flicker (*Colaptes auratus*), a large woodpecker that often feeds on the ground.

30. Grimsley and Miller, *The Union Must Stand,* 89.

31. Looby, *The Complete Civil War Journal and Selected Letters of Thomas Wentworth Higginson,* 72.

32. Quaife, *From the Cannon's Mouth,* 273.

33. Angle, *Three Years in the Army of the Cumberland,* 208.

34. Davis, *Diary of a Confederate Soldier,* 115.

35. Quaife, *From the Cannon's Mouth,* 187.

36. Rosenblatt, *Hard Marching Every Day,* 30.

37. Looby, *The Complete Civil War Journal and Selected Letters of Thomas Wentworth Higginson,* 126. Then as now, many people confused whip-poor-wills with the closely related chuck-will's-widow (*Caprimulgus carolinensis*). This soldier obviously did not.

38. Raab, *With the 3rd Wisconsin Badgers,* 195.

39. Davis, *Diary of a Confederate Soldier,* 112.

40. Snell, *Myra Inman,* 225. In this account "snowbird" may be a reference to the dark-eyed junco (*Junco hyemalis*).

41. Looby, *The Complete Civil War Journal and Selected Letters of Thomas Wentworth Higginson,* 161.

42. Patch, *This from George,* 184.

43. Westervelt and Palladino, *Diary of a Yankee Engineer,* 133.

44. Bradley, Bradley Papers, 1859–1887.

45. Burlingame and Ettlinger, *Inside Lincoln's White House,* 176.

46. Crawford, *William Howard Russell's Civil War,* 188. Colibri is the Spanish word for hummingbird.

47. Howland and Bacon, *My Heart Toward Home,* 128.

48. "Passenger Pigeon," *Encyclopedia Smithsonian.*

49. Grimsley and Miller, *The Union Must Stand,* 21.

50. Snell, *Myra Inman,* 185.

51. Holcomb, *Southern Sons, Northern Soldiers,* 12–13.

52. Garrison and Garrison, *The Encyclopedia of Civil War Usage,* 176–77.

53. *Lycoming Gazette,* April 24, 1861.

54. Westervelt and Palladino, *Diary of a Yankee Engineer,* 2.

55. McGreevy et al., *It Has Seamed Like War Today,* 10.

56. "Wadley Diary," Vol. 2, 8.

57. Clark, *Downing's Civil War Diary,* 128.

58. Holland, *Keep All My Letters,* 27.

59. Looby, *The Complete Civil War Journal and Selected Letters of Thomas Wentworth Higginson*, 121.

60. Greiner et al., *A Surgeon's Civil War*, 171.

BODY LICE, TICKS, AND HARVEST MITES

1. Billings, *Hardtack and Coffee*, 80–82.

2. Garrison and Garrison, *The Encyclopedia of Civil War Usage*, 150.

3. Miller, *Fresh Fish*, 46.

4. Brumgardt, *Civil War Nurse*, 115–16.

5. De Wolfe, *Touched with Fire*, 53.

6. Follett, letter of June 7, 1863.

7. Reinhart, *Two Germans in the Civil War*, 143–44.

8. Basile, *The Civil War Diary of Amos E. Stearns*, 103.

9. Lane, *Dear Bet*, 65.

10. Barrett, *Yankee Rebel*, 84.

11. York, *The Civil War Letters of Galutia York*, Nov. 20, 1862.

12. Sears, *On Campaign with the Army of the Potomac*, 254–55.

13. Reinhart, *Two Germans in the Civil War*, 137.

14. Andersen, *The Civil War Diary of Allen Morgan Geer*, 32.

15. Sears, *For Country, Cause & Leader*, 238.

16. Samito, *Fear Was Not in Him*, 60.

17. Westervelt and Palladino, *Diary of a Yankee Engineer*, 7–8.

18. Grimsley and Miller, *The Union Must Stand*, 221.

19. Winther, *With Sherman to the Sea*, 59.

20. Sears, *For Country, Cause & Leader*, 336.

21. Snell, *Myra Inman*, 208.

DOLPHINS, PORPOISES, AND WHALES

1. New Bedford Whaling Museum, "Overview of American Whaling."

2. Gaines, *Encyclopedia of Civil War Shipwrecks*, 17, 20–21.

3. Burlingame and Ettlinger, *Inside Lincoln's White House*, 164.

4. Winther, *With Sherman to the Sea*, 146.

5. Follett, letter of Dec. 13, 1863.

6. Rankin, *Diary of a Christian Soldier*, 89.

7. Looby, *The Complete Civil War Journal and Selected Letters of Thomas Wentworth Higginson*, 43.

8. Holcomb, *Southern Sons, Northern Soldiers*, 90.

9. Westervelt and Palladino, *Diary of a Yankee Engineer*, 108.

10. Burlingame and Ettlinger, *Inside Lincoln's White House*, 48.

11. Garrison and Garrison, *The Encyclopedia of Civil War Usage*, 235.

FISH

1. Davis, *Diary of a Confederate Soldier,* 77.
2. Oeffinger, *A Soldier's General,* 175.
3. Angle, *Three Years in the Army of the Cumberland,* 76–77.
4. Greiner et al., *A Surgeon's Civil War,* 171.
5. Grimsley and Miller, *The Union Must Stand,* 84.
6. Silver, *A Life for the Confederacy,* 146.
7. Greene, *The Civil War Diary of Lieutenant Robert Molford Addison,* 24.
8. York, The Civil War Letters of Galutia York, March 12, 1863. Speckled trout or more properly weakfish (*Cynoscion nebulosus*) are in the drum family and not related to freshwater trout.
9. Gould, *Diary of a Contraband,* 194.
10. Lane, *Dear Bet,* 51. "Stone Roller" may be a reference to the stoneroller (*Campostoma anomalum*) although this species rarely exceeds eight inches in length and is not usually considered a food fish.
11. Radigan, *Desolating This Fair Country,* 102. The American eel (*Anguilla rostrata*) is a catadromous fish, spending most of its life in fresh water but returning to the Sargasso Sea in the Atlantic Ocean to spawn.
12. Yacovone, *A Voice of Thunder,* 193.
13. Crawford, *William Howard Russell's Civil War,* 27.
14. Anderson, *Brokenburn,* 115.
15. Chesnut, *A Diary from Dixie,* 228.
16. Cumberworth and Biles, *An Enduring Love,* 164.
17. Burlingame and Ettlinger, *Inside Lincoln's White House,* 53.
18. Calkins, "A Geographic Description of the Petersburg Battlefields, June 1864–April 1865," 48–49.
19. Holland, *Keep All My Letters,* 82.
20. Peel, Civil War Diary, May 18, 1863.
21. Burlingame and Ettlinger, *Inside Lincoln's White House,* 189.
22. Lee, letter of March 29, 1983.
23. Sauers, *The Civil War Journal of Colonel William J. Bolton,* 46.
24. Kiper, *Dear Catharine, Dear Taylor,* 346.
25. Miers, *A Rebel War Clerk's Diary,* 364.
26. Ibid., 524.
27. Patch, *This from George,* 26.
28. Pelka, *The Civil War Letters of Colonel Charles F. Johnson, Invalid Corps,* 233.
29. Jackson, *Some of the Boys,* 109.
30. Grimsley and Miller, *The Union Must Stand,* 123.
31. Greene, *The Civil War Diary of Lieutenant Robert Molford Addison,* 77.
32. Sears, *For Country, Cause & Leader,* 329.

33. Clark, *Downing's Civil War Diary*, 162.

34. Jackson, *Some of the Boys*, 197.

35. Winther, *With Sherman to the Sea*, 60.

36. Northern, *All Right Let Them Come*, 90.

37. "Wadley Diary," 67.

38. Westervelt and Palladino, *Diary of a Yankee Engineer*, 26.

39. Earp, *Yellow Flag*, 44.

40. Rankin, *Diary of a Christian Soldier*, 89.

41. Ibid., 150.

42. Earp, *Yellow Flag*, 30. Nautiluses comprise a group of marine cephalopods with shells and tentacles, and are often considered "living fossils."

43. Miller, *Fresh Fish*, 43.

FLIES AND MOSQUITOES

1. "Domestic Flies," Ohio State University Extension Fact Sheet.

2. Looby, *The Complete Civil War Journal and Selected Letters of Thomas Wentworth Higginson*, 132.

3. Cutrer and Parrish, *Brothers in Gray*, 22.

4. Holcomb, *Southern Sons, Northern Soldiers*, 74–75.

5. Rosenblatt, *Hard Marching Every Day*, 132–33.

6. Grimsley and Miller, *The Union Must Stand*, 52.

7. Radigan, *Desolating This Fair Country*, 138.

8. Holcomb, *Southern Sons, Northern Soldiers*, 82.

9. Follett, letter of June 19, 1864.

10. Pelka, *The Civil War Letters of Colonel Charles F. Johnson, Invalid Corps*, 129.

11. OR, Ser. 2, Vol. 8, 605.

12. Durkin, *John Dooley Confederate Soldier*, 126.

13. Follett, letter of Sept. 24, 1863.

14. Cash and Howorth, *My Dear Nellie*, 61.

15. Jackson, *Some of the Boys*, 184.

16. Grimsley and Miller, *The Union Must Stand*, 109.

17. Sauers, *The Civil War Journal of Colonel William J. Bolton*, 16.

18. Stauffer, *Civil War Diary*. Pages unnumbered.

19. Pelka, *The Civil War Letters of Colonel Charles F. Johnson, Invalid Corps*, 161.

20. Oeffinger, *A Soldier's General*, 100. Tar was used to ward off mosquitoes.

21. Durham, *The Blues in Gray*, 34.

22. Andersen, *The Civil War Diary of Allen Morgan Geer*, 83.

23. Holcomb, *Southern Sons, Northern Soldiers*, 88.

24. Patch, *This from George*, 96.

25. Davis, *Diary of a Confederate Soldier*, 50.

26. Northern, *All Right Let Them Come*, 90.

FROGS

1. Kiper, *Dear Catharine, Dear Taylor,* 346.
2. Davis, *Diary of a Confederate Soldier,* 159.
3. Jackson, *Some of the Boys,* 184.
4. McKee, *The McKee Letters, 1859–1880,* 65.
5. Furry, *The Preacher's Tale,* 75.
6. Franklin, *The Diary of James T. Ayers,* 75.
7. York, The Civil War Letters of Galutia York, Feb. 25, 1863.
8. Basile, *The Civil War Diary of Amos E. Stearns,* 39.
9. Stauffer, *Civil War Diary.* Pages unnumbered.
10. McDonald, *Make Me a Map of the Valley,* 119.
11. Moseley, *The Stilwell Letters,* 18.

HONEYBEES

1. *Beekeeping in the United States.* USDA.
2. Ibid.
3. Porcher, 500.
4. Jacobs, "Some of the Drug Conditions During the War Between the States, 1861–5," *Southern Historical Society Papers.*
5. Hague, *A Blockaded Family,* 104.
6. Garrison and Garrison, *The Encyclopedia of Civil War Usage,* 30.
7. Robertson Jr., *Soldier of Southwestern Virginia,* 85.
8. Winther, *With Sherman to the Sea,* 135.
9. Montgomery, *Georgia Sharpshooter,* 18.
10. Davis, *Diary of a Confederate Soldier,* 52.
11. Runge, *Four Years in the Confederate Artillery,* 56.
12. Snell, *Myra Inman,* 284.
13. Child, *Letters from a Civil War Surgeon,* 279.
14. Clark, *Downing's Civil War Diary,* 261.
15. Lincecum et al., *Gideon Lincecum's Sword,* 270. The eastern kingbird *(Tyrannus tyrannus),* a type of flycatcher often called a bee martin, commonly forages on honeybees near hives.

LIZARDS

1. Berlin, *A Confederate Nurse,* 132.
2. Holcomb, *Southern Sons, Northern Soldiers,* 135.
3. Larimer, *Love and Valor,* 284.
4. Bradbury et al., *While Father Is Away,* 166.

MOLLUSKS

1. Follett, letter of Dec. 13, 1863.
2. Sauers, *The Civil War Journal of Colonel William J. Bolton,* 45.

3. Greene, *The Civil War Diary of Lieutenant Robert Molford Addison,* 23.

4. Westervelt and Palladino, *Diary of a Yankee Engineer,* 9.

5. Durham, *The Blues in Gray,* 25.

6. Speer, *Voices from Cemetery Hill,* 77.

7. Byrne, *The View from Headquarters,* 85.

OPOSSUM

1. Garrison and Garrison, *The Encyclopedia of Civil War Usage,* 195.

2. Sauers, *The Civil War Journal of Colonel William J. Bolton,* 20.

3. Looby, *The Complete Civil War Journal and Selected Letters of Thomas Wentworth Higginson,* 177.

4. Montgomery, *Georgia Sharpshooter,* 31.

5. Follett, letter of Feb. 3, 1863.

6. Looby, *The Complete Civil War Journal and Selected Letters of Thomas Wentworth Higginson,* 53.

7. Civil War Journal from the 4th Iowa Cavalry, Feb. 4, 1863.

8. Miers, *A Rebel War Clerk's Diary,* 152.

9. Ibid., 309.

OYSTERS

1. Elliot, *Doctor Quintard, Chaplain C.S.A. and Second Bishop of Tennessee,* 205.

2. Roth, *Well Mary,* 113–14.

3. Holcomb, *Southern Sons, Northern Soldiers,* 123–24.

4. Cumming, *Kate,* 248.

5. Burlingame and Ettlinger, *Inside Lincoln's White House,* 246.

6. Kiper, *Dear Catharine, Dear Taylor,* 179.

7. Gould, *Diary of a Contraband,* 222.

8. Winther, *With Sherman to the Sea,* 148.

9. Child, *Letters from a Civil War Surgeon,* 187.

10. Andersen, *The Civil War Diary of Allen Morgan Geer,* 188.

11. McKee, *The McKee Letters, 1859–1880,* 53.

12. Northern, *All Right Let Them Come,* 43.

13. Stauffer, *Civil War Diary.* Pages unnumbered.

14. Garrison and Garrison, *The Encyclopedia of Civil War Usage,* 184.

RABBITS

1. Montgomery, *Georgia Sharpshooter,* 45.

2. Northern, *All Right Let Them Come,* 68.

3. Westervelt and Palladino, *Diary of a Yankee Engineer,* 214–15.

4. Grimsley and Miller, *The Union Must Stand,* 20.

5. Sears, *On Campaign with the Army of the Potomac,* 104.

6. Smith and Baker, *Burning Rails as We Pleased,* 33.

7. Wiley, *This Infernal War*, 39.

8. Winther, *With Sherman to the Sea*, 45–46.

9. Durkin, *John Dooley Confederate Soldier*, 62.

10. Kiper, *Dear Catharine, Dear Taylor*, 179.

11. Rosenblatt, *Hard Marching Every Day*, 250.

RATS

1. Grzimek, *Animal Life Encyclopedia*.

2. Schwartz and Schwartz, *The Wild Mammals of Missouri*, 231.

3. Cunningham, *Doctors in Gray*, 177.

4. Durkin, *John Dooley Confederate Soldier*, 163.

5. Barrett, *Yankee Rebel*, 194–95.

6. Rankin, *Diary of a Christian Soldier*, 170.

7. Smith, George W. Smith Papers.

8. Bauer, *Soldiering*, 93–94.

9. Eakin and Peoples, *In Defense of My Country*, 70.

10. Snell, *Myra Inman*, 272.

11. Elliot, *Doctor Quintard, Chaplain C.S.A. and Second Bishop of Tennessee*, 175.

12. Miers, *A Rebel War Clerk's Diary*, 164.

13. Howland and Bacon, *My Heart Toward Home*, 144.

SNAKES

1. Adams, *Everybody's Magazine*, Vol. 23, No. 4.

2. Lambert, Personal Communication, 17 April 1994.

3. Wiley, *This Infernal War*, 72.

4. Winther, *With Sherman to the Sea*, 66. The rattlesnake in this anecdote is likely the cane-brake rattlesnake (*Crotalus horridus atricaudatus*), a subspecies of the timber rattlesnake.

5. Anderson, *Brokenburn*, 238.

6. Follett, letter of April 25, 1862.

7. Sears, *For Country, Cause & Leader*, 336.

8. Peoples, "An Excursion Across North Louisiana," 166.

9. Looby, *The Complete Civil War Journal and Selected Letters of Thomas Wentworth Higginson*, 140. "Thundersnake" may refer to the copperhead (*Agkistrodon contortrix*) or the eastern kingsnake (*Lampropeltis getula*).

10. Rankin, *Diary of a Christian Soldier*, 90.

11. Reinhart, *Two Germans in the Civil War*, 132.

12. Jackson, *Some of the Boys*, 75.

13. Sauers, *The Civil War Journal of Colonel William J. Bolton*, 63. "Green snakes" refers to one of two species of the genus *Opheodrys*.

14. Elder, *Love Amid the Turmoil*, 191–92.

15. Clark, *Downing's Civil War Diary*, 138. This seems to be an extreme case of exaggeration as no known snake in North America approaches fifteen feet in length.

16. Boatner, *The Civil War Dictionary*, 175.

SQUIRRELS

1. Fleming, *Civil War and Reconstruction in Alabama*, 239.

2. *Meridian Mississippian*, August 29, 1863.

3. Wiley, *This Infernal War*, 392.

4. Blackett, *Thomas Morris Chester, Black Civil War Correspondent*, 122.

5. "Civil War Journal from the 4th Iowa Cavalry," Aug. 28, 1863.

6. Bauer, *A Soldier's Journey*, 21.

7. Northern, *All Right Let Them Come*, 57.

8. Snell, *Myra Inman*, 223.

9. Angle, *Three Years in the Army of the Cumberland*, 182–83.

10. Peel, Civil War Diary, Feb. 27, 1863.

11. Grimsley and Miller, *The Union Must Stand*, 21.

12. Looby, *The Complete Civil War Journal and Selected Letters of Thomas Wentworth Higginson*, 61–62.

13. McKee, *The McKee Letters, 1859–1880*, 40.

14. Menge and Shimrak, *The Civil War Notebook of Daniel Chisholm*, 100.

15. Silver, *A Life for the Confederacy*, 82.

16. Bergeron, *The Civil War Reminiscences of Major Silas T. Grisamore, C.S.A.*, 136.

17. Westervelt and Palladino, *Diary of a Yankee Engineer*, 214.

18. Pelka, *The Civil War Letters of Colonel Charles F. Johnson, Invalid Corps*, 111. "Flying Squirls" refer to the southern flying squirrel (*Glaucomys volans*).

19. Dawson, *A Confederate Girl's Diary*, 334.

TURTLES

1. Dundee and Rossman, *The Amphibians and Reptiles of Louisiana*, 188.

2. Westervelt and Palladino, *Diary of a Yankee Engineer*, 103.

3. Ibid., 140.

4. Chesnut, *A Diary from Dixie*, 367.

5. Rankin, *Diary of a Christian Soldier*, 150.

6. Ibid., 153.

7. Crawford, *William Howard Russell's Civil War*, 25.

8. Ibid., 60.

9. "Pook Turtles." Everything2.com.

MISCELLANEOUS INVERTEBRATES

1. Roth, *Well Mary*, 65.

2. Brown, *One of Cleburne's Command*, 85.

3. Coe, *Mine Eyes Have Seen the Glory,* 197.

4. Snell, *Myra Inman,* 215. This may be a reference to cicadas.

5. Sears, *On Campaign with the Army of the Potomac,* 41.

6. Anderson, *Brokenburn,* 228.

7. Crawford, *William Howard Russell's Civil War,* 82.

8. Sears, *For Country, Cause & Leader,* 27.

9. Elder, *Love Amid the Turmoil,* 110.

10. Peoples, "An Excursion Across North Louisiana," 160.

11. Holcomb, *Southern Sons, Northern Soldiers,* 93.

12. Durham, *The Blues in Gray,* 211.

13. Sauers, *The Civil War Journal of Colonel William J. Bolton,* 69.

14. Peel, Civil War Diary, Aug. 10, 1861.

15. Coggins, *Arms and Equipment of the Civil War,* 117.

16. Holcomb, *Southern Sons, Northern Soldiers,* 135.

17. Westervelt and Palladino, *Diary of a Yankee Engineer,* 21.

MISCELLANEOUS MAMMALS

1. Schwartz and Schwartz, *The Wild Mammals of Missouri,* 324.

2. Anderson, *Brokenburn,* 63.

3. Durham, *The Blues in Gray,* 27.

4. Miers, *A Rebel War Clerk's Diary,* 328.

5. Holcomb, *Southern Sons, Northern Soldiers,* 19.

6. Garrison and Garrison, *The Encyclopedia of Civil War Usage,* 37.

7. Raccoon Roughs, 6th Alabama Infantry Regiment.

8. Rankin, *Diary of a Christian Soldier,* 168.

9. Peel, Civil War Diary, Sept. 29, 1863.

10. Lane, *Dear Bet,* 23. Referring to the inaccuracies of early surveys, wits often said that frontier surveyors measured the land with a coonskin and threw in the tail for good measure.

11. Silver, *A Life for the Confederacy,* 48.

12. Ibid., 113.

13. *The Quarterly.* Richmond National Battlefield Park newsletter, 2.

14. Grimsley and Miller, *The Union Must Stand,* 109.

15. Schwartz and Schwartz, *The Wild Mammals of Missouri,* 118.

16. Lane, *Dear Bet,* 84.

17. Menge and Shimrak, *The Civil War Notebook of Daniel Chisholm,* 62.

AFTERWORD

1. Anderson and Anderson, *The Generals,* 292.

2. Williams, *Americans and Their Forests,* 106.

3. Hunter and Hunter, *Steamboats on the Western Rivers,* 266.

4. Haites and Mak, "Steam Boating on the Mississippi, 1810–1860," 56.

5. Fishlow, *American Railroads and the Transformation of the Ante-bellum Economy*, 125–29.

6. Clark, *Downing's Civil War Diary*, 97.

7. Sears, *On Campaign with the Army of the Potomac*, 218.

8. Cunningham, *Doctors in Gray*, 149.

9. Gaines, *Encyclopedia of Civil War Shipwrecks*, 47, 122.

10. Sauers, *The Civil War Journal of Colonel William J. Bolton*, 47.

11. Greiner et al., *A Surgeon's Civil War*, 190.

12. Puckett, letter to his wife, Feb. 10, 1863.

13. Kirby, "The American Civil War."

14. "Jean Louis Rodolphe Agassiz—Biographic Notes."

15. "John Bachman's Personal Background—A Brief Biography."

16. Dorman, "The Smithsonian During the Civil War."

17. Asa Gray letter to C. R. Darwin, March 31, 1862.

18. "Jared Kirtland." Ohio History Central.

19. "John Muir, A Brief Biography."

20. J. Hennessey, National Park Service.

21. J. Lewis, National Park Service.

22. OR, Ser. 1, Vol. 24, 462–64.

23. Bollet, *Civil War Medicine*, 289.

24. Anderson, *Brokenburn*, 192.

25. Jackson, *Some of the Boy*, 109.

26. Lee, *Recollections and Letters of General Robert E. Lee*, 39.

BIBLIOGRAPHY

Adams, G. W. 1996. *Doctors in Blue: The Medical History of the Union Army in the Civil War.* Baton Rouge: Louisiana State University Press.

Adams, S. H. 1910. *Everybody's Magazine.* Vol. 23, No. 4 (Oct.).

Allen, C. M., D. A. Newman, and H. H. Winters. 2002. *Trees, Shrubs, and Woody Vines of Louisiana.* Pitkin, La.: Allen's Native Ventures.

Allen, H. D. 1918. "The Paper Money of the Confederate States. With Historical Data." *The Numismatist,* American Numismatic Association, Vol. 31, January.

"*Alligator mississippiensis.*" Louisiana Alligator Advisory Council, 2008. April 8, 2009. www.alligatorfur.com/alligator/alligator.htm.

Alspaugh, G. L. Letters. J. P. Knox Family Papers, Mss. 715, Louisiana and Lower Mississippi Valley Collections, Louisiana State University Libraries, Baton Rouge.

American Museum of Natural History. "Jean Louis Rodolphe Agassiz—Biographic Notes." 27 March 1997, Sept. 13, 2009. www.research.amnh.org/ichthyology/neoich/collectors/agassiz.html.

Andersen, M. A., ed. 1977. *The Civil War Diary of Allen Morgan Geer, 20th Regiment Illinois Volunteers.* Denver: Robert C. Appleman.

Anderson, E. M. 1972. *Memories: Historical and Personal; Including the Campaigns of the First Missouri Confederate Brigade.* 2nd ed. Dayton, Ohio: Morningside.

Anderson, J. Q., ed. 1995. *Brokenburn: The Journal of Kate Stone, 1861–1868.* Baton Rouge: Louisiana State University Press.

Anderson, N. S., and D. Anderson. 1987. *The Generals: Ulysses S. Grant and Robert E. Lee.* New York: Wings Books.

Angle, P. M., ed. 1959. *Three Years in the Army of the Cumberland: The Letters and Diary of Major James A. Connolly.* Bloomington: Indiana University Press.

Barrett, J. G., ed. 1966. *Yankee Rebel: The Civil War Journal of Edmund DeWitt Patterson.* Chapel Hill: University of North Carolina Press.

Barrow, W. C. 1990. Ecology of small insectivorous birds in a bottomland hardwood forest. PhD. diss., Louisiana State University.

Basile, L., ed. 1981. *The Civil War Diary of Amos E. Stearns, a Prisoner at Andersonville.* London: Associated University Presses.

Bauer, K. G., ed. 2001. *A Soldier's Journey: The Civil War Diary of Henry C. Caldwell, Co. E, 7th Louisiana Infantry, CSA.* Baton Rouge: Le Comité des Archives de la Louisiane.

Bauer, K. J., ed. 1988. *Soldiering: The Civil War Diary of Rice C. Bull, 123rd New York Volunteer Infantry.* New York: Berkley Books.

Bearss, E. C., ed. 1972. *A Louisiana Confederate: Diary of Felix Pierre Poche.* Natchitoches: Louisiana Studies Institute, Northwestern State University of Louisiana.

Beekeeping in the United States. 1980. USDA Agricultural Handbook No. 335. Washington.

Bergeron Jr., A. W., ed. 1993. *The Civil War Reminiscences of Major Silas T. Grisamore, C.S.A.* Baton Rouge: Louisiana State University Press.

Berlin, J. V., ed. 1994. *A Confederate Nurse: The Diary of Ada W. Bacot, 1860–1863.* Columbia: University of South Carolina Press.

Billings, J. D. 1982 [1887]. *Hardtack and Coffee.* Boston: George M. Smith & Co. Reprint, Time-Life Books.

Blackett, R. J., ed. 1989. *Thomas Morris Chester, Black Civil War Correspondent: His Dispatches from the Virginia Front.* Baton Rouge: Louisiana State University Press.

Boatner, M. M., III. 1959. *The Civil War Dictionary.* New York: David McKay.

Bollet, A. J. 2002. *Civil War Medicine: Challenges and Triumphs.* Tucson: Galen Press.

Bradbury, W. H., J. C. Bohrnstedt, and K. R. Chaney. 2003. *While Father Is Away: The Civil War Letters of William H. Bradbury.* Lexington: University Press of Kentucky.

Bradley, L. D. Bradley Papers, 1859–1887. Pearce Civil War Collection, Navarro College, Corsicana, Texas.

Bradley, M. R. 2002. *It Happened in the Civil War.* Guilford, Conn.: Globe Pequot Press.

Bragg, C. L., C. D. Ross, G. A. Blaker, S. A. Jacobe, and T. P. Savas. 2007. *Never for Want of Powder: The Confederate Powder Works in Augusta, Georgia.* Columbia: University of South Carolina Press.

Brown, N. D., ed. 1980. *One of Cleburne's Command: The Civil War Reminiscences and Diary of Capt. Samuel T. Foster, Granbury's Texas Brigade, CSA.* Austin: University of Texas Press.

Brumgardt, J. R., ed. 1980. *Civil War Nurse: The Diary and Letters of Hannah Ropes.* Knoxville: University of Tennessee Press.

Burlingame, M., and J. R. T. Ettlinger, eds. 1999. *Inside Lincoln's White House: The Complete Civil War Diary of John Hay.* Carbondale: Southern Illinois University Press.

Byrne, F. L., ed. 1965. *The View from Headquarters: Civil War Letters of Harvey Reid.* Madison: State Historical Society of Wisconsin.

Calkins, C. M. 1984. "A Geographic Description of the Petersburg Battlefields, June 1864–April 1865." *Virginia Geographer* 16 (Spring–Summer): 43–56.

Carmony, N. B., ed. 1996. *The Civil War in Apacheland: Sergeant George Hand's Diary: California, Arizona, West Texas, New Mexico, 1861–1864.* Silver City, N.M.: High-Lonesome Books.

Cash, W. M., and L. S. Howorth, eds. 1977. *My Dear Nellie: The Civil War Letters of William L. Nugent to Eleanor Smith Nugent.* Oxford: University Press of Mississippi.

Cassidy, F. G., and J. H. Hall, eds. 1985. *Dictionary of American Regional English, Vol. 1.* Cambridge, Mass.: Belknap Press.

Chesnut, M. B. 1980. *A Diary from Dixie.* Ed. B. A. Williams. Cambridge: Harvard University Press.

Child, W. 1995. *Letters from a Civil War Surgeon: The Letters of Dr. William Child of the Fifth New Hampshire Volunteers.* Transcribed by M. C. Sawyer, B. Sawyer, and T. C. Sawyer. Solon, Maine: Polar Bear.

"Cinchona." Tropical Plant Database. Jan. 11, 2008. www.rain-tree.com/quinine.htm.

Clark, O. B., ed. 1916. *Downing's Civil War Diary.* Des Moines: Historical Dept. of Iowa.

Coe, D., ed. 1975. *Mine Eyes Have Seen the Glory: Combat Diaries of Union Sergeant Hamlin Alexander Coe.* Cranbury, N.J.: Associated University Presses.

Coggins, J. 2004. *Arms and Equipment of the Civil War.* Needham, Mass.: Dover Press.

Confederate Receipt Book: A Compilation of Over One Hundred Receipts, Adapted to the Times. 1863. Richmond: West & Johnson.

Corbusier, W. H., and R. Wooster, ed. 2003. *Soldier, Surgeon, Scholar: The Memoirs of William Henry Corbusier.* Norman: University of Oklahoma Press.

Cotton, G. A., ed. 1984. *Yankee Bullets, Rebel Rations: Caught Between Two Armies, Vicksburg Citizens Recall the Horrors of the 1863 Siege.* Raymond, Miss.: Keith.

Cowdrey, A. E. 1983. *This Land, This South: An Environmental History.* Lexington: University Press of Kentucky.

Craig, W. S. "The Civil War Letters of William S. Craig—Sept. 8, 1863." April 6, 2009. www.ehistory.osu.edu/osu/sources/letters/craig/004.cfm.

Craughwell, T. J. 2008. *Stealing Lincoln's Body.* Cambridge, Mass.: Belknap Press.

Crawford, M., ed. 1992. *William Howard Russell's Civil War: Private Diary and Letters, 1861–1862.* Athens: University of Georgia Press.

Cullina, W. 2000. *The New England Wild Flower Society Guide to Growing and Propagating Wildflowers of the United States and Canada.* New York: Houghton Mifflin Harcourt.

Cumberworth, S. M., and D. V. Biles, eds. 1995. *An Enduring Love: The Civil War Diaries of Benjamin Franklin Pierce (14th New Hampshire Vol. Inf.) and His Wife Harriett Jane Goodwin Pierce.* Gettysburg, Penn.: Thomas Publications.

Cumming, Kate. 1998. *Kate: The Journal of a Confederate Nurse.* Ed. R. B. Harwell. Baton Rouge: Louisiana State University Press.

Cunningham, H. H. 1958. *Doctors in Gray: The Confederate Medical Service.* Baton Rouge: Louisiana State University Press.

Cutrer, T. W., ed. 2002. *Oh, What a Loansome Time I Had: The Civil War Letters of Major William Morel Moxley, Eighteenth Alabama Infantry, and Emily Beck Moxley.* Tuscaloosa: University of Alabama Press.

Cutrer, T. W., and T. M. Parrish, eds. 1997. *Brothers in Gray: The Civil War Letters of the Pierson Family.* Baton Rouge: Louisiana State University Press.

Davis, W. C., ed. 1990. *Diary of a Confederate Soldier: John S. Jackman of the Orphan Brigade.* Columbia: University of South Carolina Press.

Dawson, S. M. 1913. *A Confederate Girl's Diary.* Boston: Houghton Mifflin.

Dennis, F. A., ed. 1973. *Kemper County Rebel: The Civil War Diary of Robert Masten Holmes, C.S.A.* Jackson: University and College Press of Mississippi.

De Wolfe, M., ed. 1947. *Touched with Fire: Civil War Letters and Diary of Oliver Wendell Holmes, Jr. 1861–1864.* Cambridge, Mass.: Harvard University Press.

Dorman, K. W. "The Smithsonian During the Civil War." The Smithsonian Associates Civil War E-Mail Newsletter, Vol. 5, No. 9, Sept. 13, 2009. www .civilwarstudies.org/articles/Vol_5/si-during-civil-war.shtm 9/13/2009.

Dundee, H. A., and D. A. Rossman. 1989. *The Amphibians and Reptiles of Louisiana.* Baton Rouge: Louisiana State University Press.

Durham, R. S., ed. 2000. *The Blues in Gray: The Civil War Journal of William Daniel Dixon and The Republican Blues Daybook.* Knoxville: University of Tennessee Press.

Durkin, J. T., ed. 1963. *John Dooley, Confederate Soldier: His War Journal.* Notre Dame, Ind.: University of Notre Dame Press.

Eakin, S. L., and M. Peoples, eds. 1983. *"In Defense of My Country . . .": The Letters of a Shiloh Confederate Soldier, Sergeant George Washington Bolton, and His Union Parish Neighbors of the Twelfth Regiment of Louisiana Volunteers (1861–1864).* Pub. for the Corney Creek Festival.

Earley, L. S. 2004. *Looking for Longleaf: The Fall and Rise of an American Forest.* Chapel Hill: University of North Carolina Press.

Earp, C. A., ed. 2002. *Yellow Flag: The Civil War Journal of Surgeon's Steward C. Marion Dodson.* Baltimore: Maryland Historical Society.

Edlin, H. L. 1978. *The Tree Key.* New York: Charles Scribner's Sons.

Elder, D. C., III, ed. 2003. *Love Amid the Turmoil: The Civil War Letters of William and Mary Vermilion.* Iowa City: University of Iowa Press.

Elliot, S. D., ed. 2003. *Doctor Quintard, Chaplain C.S.A. and Second Bishop of Tennessee: The Memoir and Civil War Diary of Charles Todd Quintard.* Baton Rouge: Louisiana State University Press.

Etnier, D. A. 1997. "Jeopardized Southeastern Freshwater Fishes: A Search for Causes." In G. W. Benz and D. E. Collins, eds., *Aquatic Fauna in Peril—The Southeastern Perspective.* Special Publication 1. 87–104. Decatur, Ga.: Southeast Aquatic Research Institute.

Fehrenbacher, D. E., ed. 1989. *Abraham Lincoln: Speeches and Writings, 1859–1865: Speeches, Letters, and Miscellaneous Writings, Presidential Messages and Proclamations.* New York: Library of America.

Fishlow, A. 1965. *American Railroads and the Transformation of the Ante-bellum Economy.* Harvard Economic Studies 127. Cambridge: Harvard University Press.

Flannery, M. A. 2004. *Civil War Pharmacy: A History of Drugs, Drug Supply and Provision, and Therapeutics for the Union and Confederacy.* New York: Pharmaceutical Products Press.

Fleming, W. L. 1905. *Civil War and Reconstruction in Alabama.* New York: Columbia University Press.

Follett, J. M. Letter of April 25, 1862. April 8, 2009. www.ehistory.osu.edu/osu/sources/letters/follett_brothers/letters/JF620425.cfm.

———. Letter of April 5, 1863. April 8, 2009. www.ehistory.osu.edu/osu/sources/letters/follett_brothers/letters/JF630405.cfm.

———. Letter of June 7, 1863. April 8, 2009. www.ehistory.osu.edu/osu/sources/letters/follett_brothers/letters/JF630607.cfm.

———. Letter of Dec. 13, 1863. April 8, 2009. www.ehistory.osu.edu/osu/sources/letters/follett_brothers/letters/JF631213.cfm.

———. Letter of June 19, 1864. April 8, 2009. www.ehistory.osu.edu/osu/sources/letters/follett_brothers/letters/JF640619.cfm.

Follett, M. C. Letter of Feb. 3, 1863. April 8, 2009. www.ehistory.osu.edu/osu/sources/letters/follett_melville/pages/follett_diary8n9.cfm.

———. Letter of Feb. 8, 1863. April 8, 2009. www.ehistory.osu.edu/osu/sources/letters/follett_melville/pages/follett_diary12n13.cfm.

———. Letter of Sept. 24, 1863. April 8, 2009. www.ehistory.osu.edu/osu/sources/letters/follett_melville/pages/follett_diary60n61.cfm.

Foster, S., and V. E. Tyler. 2000. *Tyler's Honest Herbal—A Sensible Guide to the Use of Herbs and Related Remedies.* New York: Haworth Herbal Press.

Franklin, J. H., ed. 1999. *The Diary of James T. Ayers, Civil War Recruiter.* Baton Rouge: Louisiana State University Press.

Furry, W., ed. 2001. *The Preacher's Tale: The Civil War Journal of Rev. Francis Springer, Chaplain, U.S. Army of the Frontier.* Fayetteville: University of Arkansas Press.

Gabrielson, I. N. 1951. *Wildlife Management.* New York: Macmillan.

Gaines, W. C. 2008. *Encyclopedia of Civil War Shipwrecks.* Baton Rouge: Louisiana State University Press.

Galbraith, L. and W., eds. 1990. *A Lost Heroine of the Confederacy: The Diaries and Letters of Belle Edmondson.* Oxford: University Press of Mississippi.

Garrison, W., with C. Garrison. 2001. *The Encyclopedia of Civil War Usage: An Illustrated Compendium of the Everyday Language of Soldiers and Civilians.* Nashville: Cumberland House.

Gettysburg National Military Park. News Release, Aug. 18, 2008. Feb. 16, 2009. www.friendsofgettysburg.org/UpdateonGettysburgsWitnessTree.htm.

Goodwin, G. G. 1936. "Big Game Animals in the Northeastern United States." *Journal of Mammalogy* 17:48–50.

Gould, W. B., IV, ed. 2002. *Diary of a Contraband: The Civil War Passage of a Black Sailor.* Stanford, Calif.: Stanford University Press.

Gray, Asa, to C. R. Darwin. Letter 3489. March 31, 1862. Darwin Correspondence Project. Sept. 13, 2009. www.darwinproject.ac.uk/darwinletters/calendar/entry-3489.html.

Greene, D. E., ed. 2001. *The Civil War Diary of Lieutenant Robert Molford Addison, Co. E, 23rd Wisconsin Infantry.* Westminster, Md.: Willow Bend Books.

Greiner, J. M., J. L. Coryell, and J. R. Smither, eds. 1994. *A Surgeon's Civil War: The Letters and Diary of Daniel M. Holt, M.D.* Kent, Ohio: Kent State University Press.

Grimsley, M., and T. D. Miller, eds. 2000. *The Union Must Stand: The Civil War Diary of John Quincy Adams Campbell, Fifth Iowa Volunteer Infantry.* Knoxville: University of Tennessee Press.

Grzimek, B. 1968. *Animal Life Encyclopedia.* New York: Van Nostrand Reinhold Co.

Hague, P. A. 1888. *A Blockaded Family: Life in Southern Alabama During the Civil War.* Boston: Houghton, Mifflin.

Haites, E. F., and J. Mak. 1971. "Steam Boating on the Mississippi, 1810–1860: A Purely Competitive Industry." *Business History Review* 45, 52–79.

Hammer, J. J., ed. 1973. *Frederic Augustus James's Civil War Diary.* Madison, N.J.: Fairleigh Dickinson University Press.

Harrar, E. S., and J. G. Harrar. 1962. *Guide to Southern Trees.* New York: Dover.

Hart, A. B., and E. Stevens. 1903. *The Romance of the Civil War.* New York: Macmillan.

Heidler, D., J. T. Heidler, and D. J. Coles, eds. 2002. *Encyclopedia of the American*

Civil War: A Political, Social and Military History. Scranton, Penn.: W. W. Norton.

Hennessey, J. National Park Service. Personal Communication, 6 July 2006.

Higginson, T. W. 1870. *Army Life in a Black Regiment.* Boston: Fields, Osgood & Co.

Holcomb, J., ed. 2004. *Southern Sons, Northern Soldiers: The Civil War Letters of the Remley Brothers, 22nd Iowa Infantry.* DeKalb: Northern Illinois University Press.

Holland, K. S., ed. 2003. *Keep All My Letters: The Civil War Letters of Richard Henry Brooks, 51st Georgia Infantry.* Macon, Ga.: Mercer University Press.

Howland, E. W., and G. W. Bacon. 2001. *My Heart Toward Home: Letters of a Family During the Civil War.* Ed. D. J. Hoisington. Roseville, Minn.: Edinborough Press.

Hubbard, W. F. 1904. *The Basket Willow.* U.S. Department of Agriculture Bureau of Forestry, Bull. No. 46.

Hunter, L. C., and B. J. Hunter. 1994. *Steamboats on the Western Rivers: An Economic and Technological History.* North Chelmsford, Mass.: Courier Dover.

Husby, K. J., and E. J. Wittenberg, eds. 2000. *Under Custer's Command: The Civil War Journal of James Henry Avery.* London: Brassey's.

Iowa, 4th Iowa Cavalry. "Civil War Journal from the 4th Iowa Cavalry." Feb. 4, 1863. Feb. 11, 2009. www.garthhagerman.com/fambly/bereman2.php.

———. May 19, 1863. Feb. 11, 2009. www.garthhagerman.com/fambly/bereman4 .php.

———. July 25, 1863. Feb. 11, 2009. www.garthhagerman.com/fambly/bereman7 .php.

———. Aug. 28, 1863. Feb. 11, 2009. www.garthhagerman.com/fambly/bereman8 .php.

Irion, J. B., D. V. Beard, and P. V. Heinrich. 1994. *Remote Sensing Investigations of Civil War Era Shipwrecks in the Vicinity of Fort St. Philip, Plaquemines Parish, Louisiana.* New Orleans: Goodwin and Associates.

Jackson, J. O., ed. 1960. *"Some of the Boys . . .": The Civil War Letters of Isaac Jackson 1862–1865.* Carbondale: Southern Illinois University Press.

Jacobs, J. 1905. "Some of the Drug Conditions During the War Between the States, 1861–5." *Southern Historical Society Papers.* Vol. 33, Jan.–Dec. Feb. 10, 2009. www.civilwarhome.com/drugsshsp.htm.

"Jared Kirtland." Ohio History Central. Sept. 13, 2009. www.ohiohistorycentral .org/entry.php?rec=229.

"John Bachman's Personal Background—A Brief Biography." Newberry College Alumni Association. Sept. 13, 2009. www.johnbachman.org/HPersonal BackgroundMain.html.

"John Muir: A Brief Biography." Ecology Hall of Fame. July 12, 1997, Sept. 13, 2009. www.ecotopia.org/ehof/muir/bio.html.

Johnson, R. U., and C. C. Buel, eds. 1887–1888. *Battles and Leaders of the Civil War.* 4 vols. New York.

Jones, T. L., ed. 1991. *The Civil War Memoirs of Captain William J. Seymour: Reminiscences of a Louisiana Tiger.* Baton Rouge: Louisiana State University Press.

Kiper, R. L., ed. 2002. *Dear Catharine, Dear Taylor: The Civil War Letters of a Union Soldier and His Wife.* Lawrence: University Press of Kansas.

Kirby, J. T. "The American Civil War: An Environmental View." National Humanities Center TeacherServe Website. July 2001, Aug. 25, 2009. www.nationalhumanitiescenter.org/tserve/nattrans/ntuseland/essays/amcwar.htm.

Konstam, A. 2002. *Mississippi River Gunboats of the American Civil War, 1861–1865.* Oxford: Osprey.

Lambert, J. M. M. Personal Communication, April 11, 1994.

Lane, B. M., ed. 1978. *Dear Bet: The Carter Letters, 1861–1863.* Self-published.

Larimer, C. F., ed. 2000. *Love and Valor: The Intimate Civil War Letters Between Captain Jacob and Emeline Ritner.* Chicago: Sigourney Press.

The Leather Manufacturer. 1898. Vol. 9, No. 1. Author unknown. April 8, 2009. www.books.google.com.

Lee, R. E. Letter of March 29, 1863. Washington and Lee University. April 9, 2009. www.home.wlu.edu/~stanleyv/pentrans.htm.

Lee Jr., R. E. 1904. *Recollections and Letters of General Robert E. Lee.* Garden City, N.Y.: Garden City Publishing.

Lewis, J. National Park Service. Personal Communication, July 10, 2006.

Lincecum, J. B., E. H. Phillips, and P. A. Redshaw, eds. 2001. *Gideon Lincecum's Sword: Civil War Letters from the Texas Home Front.* Denton: University of North Texas Press.

Lock, S., J. M. Last, and G. Dunea, eds. 2001. *The Oxford Illustrated Companion to Medicine.* New York: Oxford University Press.

Looby, C., ed. 2000. *The Complete Civil War Journal and Selected Letters of Thomas Wentworth Higginson.* Chicago: University of Chicago Press.

Lowe, R., ed. 1999. *A Texas Cavalry Officer's Civil War: The Diary and Letters of James C. Bates.* Baton Rouge: Louisiana State University Press.

Lycoming Gazette. April 24, 1861. Williamsport, Pennsylvania.

"Magnolia Hall." Feb. 9, 2009. www.natchezontheriver.com/news/2008/oct/13/magnolia-hall.

Manarin, L. 1979. *North Carolina Troops 1861–1865: Infantry 22nd–26th Regiments, Vol. VII.* North Carolina Historical Publications Section.

"Maple Sugaring History, Naper Settlement." Feb. 7, 2009. www.napersettlement.org/calendar/Maple%20Sugaring%20Days.html.

Martin, A. C., H. S. Zim, and A. L. Nelson. 1961. *American Wildlife & Plants—A Guide to Wildlife Food Habits.* New York: Dover.

Martin, W. H., S. G. Boyce, and A. C. Echternacht, eds. 1993. *Biodiversity of the*

Southeastern United States—Lowland Terrestrial Communities. New York: John Wiley and Sons.

Maryland State Archives. "Mills." Feb. 9, 2009. www.msa.md.gov/msa/refserv/html/mills.html.

Massey, M. E. 1952. *Ersatz in the Confederacy.* Columbia: University of South Carolina Press.

Maynard, D., ed. 1964. "Vicksburg Diary: The Journal of Gaberiel M. Killgore." *Civil War History* 10, no. 1:33–53.

McDonald, A. P., ed. 1989. *Make Me a Map of the Valley: The Civil War Journal of Stonewall Jackson's Topographer.* Dallas: Southern Methodist University Press.

McGreevy, C., M. A. Mogus, and C. Murphy, eds. 1999. *"It Has Seamed Like War Today": The Civil War Letters of William D. Walton.* Stroudsburg, Penn.: Monroe County Historical Assoc.

McGuire, H. H. 1889. "Progress of Medicine in the South." *Southern Historical Society Papers,* Vol. 17.

McKee, H., ed. 2000. *The McKee Letters, 1859–1880: Correspondence of a Georgia Farm Family During the Civil War and Reconstruction.* Milledgeville, Ga.: Boyd.

Menge, W. S., and J. A. Shimrak, eds. 1989. *The Civil War Notebook of Daniel Chisholm: A Chronicle of Daily Life in the Union Army, 1864–1865.* London: Orion Books.

Meridian Mississippian, August 29, 1863.

Miers, E. S., ed. 1958. *A Rebel War Clerk's Diary.* New York: Sagamore.

Miller, C. D., S. Bennett, and B. Tillery. 2004. *The Struggle for the Life of the Republic: A Civil War Narrative.* Kent, Ohio: Kent State University Press.

Miller, E. L. 2003. *Fresh Fish: A Civil War Prisoner's Story.* Parsons, W.Va.: McClain.

Montgomery Jr., G. F., ed. 1997. *Georgia Sharpshooter: The Civil War Diary and Letters of William Rhadamanthus Montgomery, 1839–1906.* Macon: Mercer University Press.

Moore, J. H. 1967. *Andrew Brown and Cypress Lumbering in the Old Southwest.* Baton Rouge: Louisiana State University Press.

Moore, S. P. 1863. "Standard Supply Table of the Indigenous Remedies for Field Service and the Sick in General Hospitals." Richmond: Surgeon General's Office.

Moseley, R. H., ed. 2002. *The Stilwell Letters: A Georgian in Longstreet's Corps, Army of Northern Virginia.* Macon: Mercer University Press.

Nash, H. M. "Some Reminiscences of a Confederate Surgeon." *Transactions of the College of Physicians of Philadelphia,* third series, 28 (1906): 133.

National Park Service. 2007. *Historical Land Cover/Use Classification of Pea Ridge National Military Park.* National Park Service.

———. "*Maple Leaf* Shipwreck." Feb. 7, 2009. www.nps.gov/history/NR/travel/ flshipwrecks/map.htm.

National Pecan Sheller's Association. "History of Pecans." April 16, 2009. http:// www.ilovepecans.org/history.html.

New Bedford Whaling Museum. "Overview of American Whaling." Jan. 25, 2009. www.whalingmuseum.org/library/amwhale/am_index.html.

Northern, C. S., III, ed. 2003. *All Right Let Them Come: The Civil War Diary of an East Tennessee Confederate.* Knoxville: University of Tennessee Press.

Oeffinger, J. C., ed. 2002. *A Soldier's General: The Civil War Letters of Major General Lafayette McLaws.* Chapel Hill: University of North Carolina Press.

Official Records of the Union and Confederate Navies in the War of the Rebellion. 1894– 1922. 31 vols. Washington.

Ohio State University Extension. "Domestic Flies." Ohio State University Extension Fact Sheet. Jan. 12, 2008. www.ohioline.osu.edu/hyg-fact/2000/2111 .html.

"Passenger Pigeon." *Encyclopedia Smithsonian.* 3/01. Jan. 21, 2008. www.si.edu/ Encyclopedia_SI/nmnh/passpig.htm.

Patch, E. M. K., ed. 2001. *This from George: The Civil War Letters of Sergeant George Magusta Englis, 1861–1865, Company K, 89th New York Regiment of Volunteer Infantry, known as the Dickinson Guard.* Binghamton, N.Y.: Broome County Historical Society.

Peabody, E. P., ed. 1849. *Aesthetic Papers,* 231. Boston and New York.

Peel, A. L. Civil War Diary. Aug. 10, 1861. April 6, 2009. www.myweb.cableone.net/ 4jdurham/peel/61peelaug.html.

———. Civil War Diary. Sept. 22, 1861. April 6, 2009. www.myweb.cableone.net/ 4jdurham/peel/61peelsep.html.

———. Civil War Diary. Feb. 27, 1863. April 6, 2009. www.myweb.cableone.net/ 4jdurham/peel/peeljan_april.html.

———. Civil War Diary. May 8, 1863. April 6, 2009. www.myweb.cableone.net/ 4jdurham/peel/peelmay.html.

———. Civil War Diary. May 16, 1863. April 6, 2009. www.myweb.cableone.net/ 4jdurham/peel/peelmay.html.

———. Civil War Diary. May 18, 1863. April 6, 2009. www.myweb.cableone.net/ 4jdurham/peel/peelmay.html.

———. Civil War Diary. Sept. 29, 1863. April 6, 2009. www.freepages.family .rootsweb.com/~peel/peelsept.html.

Pelka, F., ed. 2004. *The Civil War Letters of Colonel Charles F. Johnson, Invalid Corps.* Amherst: University of Massachusetts Press.

Peoples, M., ed. 1977. "An Excursion Across North Louisiana: Excerpts from the Diary of British Lieutenant Colonel Thomas Freemantle (May 8 to May 15, 1863)." *North Louisiana Historical Association Journal* 8/4: 159–69.

Peters, P. R. 2001. *The Underground Railroad in Floyd County, Indiana.* Jefferson, N.C.: McFarland.

Poe, J. C., ed. 1967. *The Raving Foe: The Civil War Diary of Major James T. Poe, C.S.A. and the 11th Arkansas Volunteers.* Longhorn Press.

"Pook Turtles." Everything2. July 20, 2001, Jan. 1, 2009. www.everything2.com/ e2node/Pook%2520Turtles.

Porcher, F. S. 1863. *Resources of the Southern Fields and Forests, Medical, Economical, and Agricultural. Being also a Medical Botany of the Confederate States: with Practical Information on the Useful Properties of the Trees, Plants, and Shrubs.* Charleston, S.C.: Steam-Power Press of Evans & Cogswell.

Porter, D. D. 1886. *The Naval History of the Civil War.* New York: Sherman.

Puckett, J. H. Letter to his wife, Feb. 10, 1863. Manuscript, University of Texas.

Quaife, M. M., ed. 1959. *From the Cannon's Mouth: The Civil War Letters of General Alpheus S. Williams.* Detroit: Wayne State University Press and the Detroit Historical Society.

Raab, S. S., ed. 1999. *With the 3rd Wisconsin Badgers: The Living Experience of the Civil War Through the Journals of Van R. Willard.* Mechanicsburg, Penn.: Stackpole Books.

Raccoon Roughs. 6th Alabama Infantry Regiment. April 10, 2009. www.rootsweb .ancestry.com/~alcwroot/6th_alabama_inf/regt_officers.htm.

Radigan, E. N., ed. 1999. *Desolating This Fair Country: The Civil War Diary and Letters of Lt. Henry C. Lyon, 34th New York.* Jefferson, N.C.: McFarland and Company.

Rankin, D. C., ed. 2004. *Diary of a Christian Soldier: Rufus Kinsley and the Civil War.* Cambridge: Cambridge University Press.

Reddick, H. W. 1910. *Seventy-Seven Years in Dixie: The Boys in Gray of 61–65.* Self-published.

Reinhart, J. R., ed. 2004. *Two Germans in the Civil War: The Diary of John Daeuble and the Letters of Gottfried Rentschler, 6th Kentucky Volunteer Infantry.* Knoxville: University of Tennessee Press.

Richmond National Battlefield Park. *The Quarterly.* Richmond National Battlefield Park newsletter, 14: July 2003.

Robertson Jr., J. I., ed. 2004. *Soldier of Southwestern Virginia: The Civil War Letters of Captain John Preston Sheffey.* Baton Rouge: Louisiana State University Press.

Robin, C. C. 1966. *Voyage to Louisiana, 1803–1805.* Trans. S. O. Landry Jr. New Orleans: Pelican.

Roosevelt, T. 1908. "In the Louisiana Canebrakes." *Scribners Magazine* 43:47–60.

"Rope-tension drums." National Music Museum. Oct. 9, 2008, Feb. 7, 2009. www .usd.edu/smm/RopeTensionSnareDrums.html#2866.

Rosenblatt, E. and R., eds. 1992. *Hard Marching Every Day: The Civil War Letters of Private Wilbur Fisk, 1861–1865.* Lawrence: University Press of Kansas.

Roth, M. B., ed. 1960. *Well Mary: Civil War Letters of a Wisconsin Volunteer.* Madison: University of Wisconsin Press.

Runge, W. H., ed. 1961. *Four Years in the Confederate Artillery: The Diary of Private Henry Robinson Berkeley.* Virginia Historical Society (by University of North Carolina Press).

Samito, C. G., ed. 2004. *"Fear Was Not in Him": The Civil War Letters of Major General Francis C. Barlow, U.S.A.* New York: Fordham University Press.

Sanders, J. Y. [1863] 2004. "Diary." In A. C. Richard Jr. and M. M. Richard, *The Defense of Vicksburg—A Louisiana Chronicle.* College Station: Texas A&M Press.

Sargent, C. S. 1905. *Manual of the Trees of North America (Exclusive of Mexico).* Boston: Houghton Mifflin.

Sauers, R. A., ed. 2000. *The Civil War Journal of Colonel William J. Bolton, 51st Pennsylvania, April 20, 1861–August 2, 1865.* Conshohocken, Penn.: Combined.

Schwartz, C. W., and E. R. Schwartz. 1971. *The Wild Mammals of Missouri.* Columbia: University of Missouri Press.

Schwartz, G., ed. 1989. *A Woman Doctor's Civil War.* Columbia: University of South Carolina Press.

Sears, S. W., ed. 1993. *For Country, Cause & Leader: The Civil War Journal of Charles B. Haydon.* Boston: Ticknor and Fields.

———, ed. 2001. *On Campaign with the Army of the Potomac: The Civil War Journal of Theodore Ayrault Dodge.* Lanham, Md.: Cooper Square Press.

Sherman, W. T. 1875. *Memoirs of General William T. Sherman.* 2 vols. New York: D. Appleton.

Silver, J. W., ed. 1959. *A Life for the Confederacy: As Recorded in the Pocket Diaries of Pvt. Robert A. Moore.* Jackson, Tenn.: McCowat-Mercer Press.

Skipper, M., and J. Taylor, eds. 2004. *A Handful of Providence: The Civil War Letters of Lt. Richard Goldwaite, New York Volunteers, and Ellen Goldwaite.* Jefferson, N.C.: McFarland.

Smith, B. B., and N. B. Baker, eds. 2004. *"Burning Rails as We Pleased": The Civil War Letters of William Garrigues Bentley, 104th Ohio Volunteer Infantry.* Jefferson, N.C.: McFarland.

Smith, G. W. 1864. George W. Smith Papers. West Virginia and Regional History Collection, West Virginia University Libraries.

Smith, J. D., and W. Cooper Jr., eds. 2000. *A Union Woman in Civil War Kentucky: The Diary of Frances Peter.* Lexington: University of Kentucky Press.

Snell, W. R., ed. 2000. *Myra Inman: A Diary of the Civil War in East Tennessee.* Macon: Mercer University Press.

Southgate, E. W. "Historical Ecology of American Chestnut (*Castanea dentata*)." Jan. 26, 2008. www.chestnut.cas.psu.edu/Meetings/NPS/proceedings/_2_03 -Southgate%20manuscript.pdf.

Speer, A. P., ed. 1997. *Voices from Cemetery Hill: The Civil War Diary, Reports, and*

Letters of Colonel William Henry Asbury Speer (1861–1864). Johnson City, Tenn.: Overmountain Press.

Stauffer, N. 1976. *Civil War Diary.* California State University, Northridge Libraries.

Stevens, W. B. 1921. *Centennial History of Missouri.* Vol. 2. St. Louis: S. J. Clark.

Still Jr., W. N. 1969. *Confederate Shipbuilding.* Athens: University of Georgia Press.

Straubing, H. E., ed. 1993. *In Hospital and Camp: The Civil War Through the Eyes of Its Doctors and Nurses.* Mechanicsburg, Penn.: Stackpole Books.

Sutherland, D. E., ed. 1996. *A Very Violent Rebel: The Civil War Diary of Ellen Renshaw House.* Knoxville: University of Tennessee Press.

Swedberg, C. E., ed. 1999. *Three Years with the 92nd Illinois: The Civil War Diary of John M. King.* Mechanicsburg, Penn.: Stackpole Books.

Taylor, F. J., ed. 1959. *Reluctant Rebel: The Secret Diary of Robert Patrick, 1861–1865.* Baton Rouge: Louisiana State University Press.

Texas Confederate Military Organizations. Feb. 1, 2009. www.tarleton.edu/~kjones/localtx.html.

Thomas, R. D. 2003. *100 Woody Plants of North Louisiana.* Monroe, La.: Herbarium of University of Louisiana at Monroe.

Trimble, R. M., ed. 2000. *Brothers 'Til Death: The Civil War Letters of William, Thomas, and Maggie Jones, 1861–1865.* Macon: Mercer University Press.

U.S. Bureau of the Census. June 15, 1998. www.census.gov/population/documentation/twps0027/tab09.txt.

U.S. Census Data. "Population of the United States—1860." April 6, 2009. www.civilwarhome.com/population1860.htm.

U.S. Forest Service. "Gypsy Moth in North America." Oct. 29, 2003, Feb. 11, 2009. www.fs.fed.us/ne/morgantown/4557/gmoth.

———. "White Mulberry." July 22, 2005, Feb. 11, 2009. www.invasive.org/weedcd/pdfs/wow/white_mulberry.pdf.

U.S. War Department. 1880–1901. *The War of the Rebellion: A Compilation of the Official Records of the Union and Confederate Armies.* 128 vols., series 1–3. Washington, D.C.

Wadley, S. L. "Sarah L. Wadley Diary, August 8, 1859–May 15 1865." Feb. 3, 2007. www.docsouth.unc.edu/imls/wadley/menu.html.

Weaver, C. P., ed. 1998. *Thank God My Regiment an African One: The Civil War Diary of Colonel Nathan W. Daniels.* Baton Rouge: Louisiana State University Press.

Westervelt, J. H., and A. Palladino, eds. 1996. *Diary of a Yankee Engineer: The Civil War Story of John H. Westervelt, Engineer, 1st New York Volunteer Engineer Corps.* Bronx, N.Y.: Fordham University Press.

Whitney, G. G. 1994. *From Coastal Wilderness to Fruited Plain: A History of Environmental Change in Temperate North America 1500 to the Present.* Cambridge: Cambridge University Press.

Wiley, B. I., ed. 1958. *"This Infernal War": The Confederate Letters of Sgt. Edwin H. Fay.* Austin: University of Texas Press.

Williams, M. 1989. *Americans and Their Forests: A Historical Geography.* Cambridge: Cambridge University Press.

Winther, O. O., ed. 1958. *With Sherman to the Sea: The Civil War Letters, Diaries & Reminiscences of Theodore F. Upson.* Bloomington: Indiana University Press.

Yacovone, D., ed. 1997. *A Voice of Thunder: The Civil War Letters of George E. Stephens.* Champaign: University of Illinois Press.

York, Galutia. The Civil War Letters of Galutia York. Nov. 20, 1862. Sept. 8, 2005, Jan. 7, 2007. www.library.morrisville.edu/local_history/civil_war/nov_20.html.

———. Jan. 7, 1863. Sept. 8, 2005, Jan. 7, 2007. www.library.morrisville.edu/local_history/civil_war/jan_7.html.

———. Feb. 1, 1863. Sept. 8, 2005, Jan. 7, 2007. www.library.morrisville.edu/local_history/civil_war/feb_1.html.

———. Feb. 14, 1863. Sept. 8, 2005, Jan 7, 2007. www.localhistory.morrisville.edu/civil_war/feb_14.html.

———. Feb. 19, 1863. Sept. 8, 2005, Jan. 7, 2007. www.localhistory.morrisville.edu/civil_war/feb_19.html.

———. Feb. 25, 1863. Sept. 8, 2005, Jan. 7, 2007. www.library.morrisville.edu/local_history/civil_war/feb_25.html.

———. March 12, 1863. Sept. 8, 2005, Jan. 7, 2007. www.library.morrisville.edu/local_history/civil_war/mar_12.html.

———. May 16, 1863, Jan. 7, 2007. www.library.morrisville.edu/local_history/civil_war/may_16.html.

INDEX